SUPPLEMENT Secrets

The Innovatively Simple

Supplement Encyclopedia

Discover the Supplements Proven to Create a Better You
That's disrupting a multi-billion industry

Jonathon Castillo

www.endolls.com
Print ISBN-13: 978-0-578-55368-9

Disclaimer

The content provided in this book is designed to provide helpful information on the subjects discussed; it is not meant to be used, nor should it be used, to diagnose or treat any medical condition. For diagnosis and treatment of any medical problem, consult with your doctor, physician, or other qualified medical professional.

All material and any content contained on or in this book is for informational purposes only and is not intended to substitute for the advice, diagnosis, or treatment from a qualified or licensed medical professional. Always consult with your doctor, physician, or other qualified medical professional with any medical questions you may have.

The publisher, the book, and the writer do not make any representations, recommend, or endorse any specific tests, products, procedures, opinions, or other information that is mentioned in or on this book. Additionally, we are not responsible for any specific health or allergy needs that may require medical supervision and are not liable for any damages or negative consequences for any treatment, action, application or preparation to any person reading or following the information in this book. Reliance on any information provided by Supplement Secrets or its writer, Jonathon Castillo, is solely at your own risk.

Dedication

I made this book after seeing all the misinformation out there. I remember walking into numerous nutrition stores, including those Big-Name Big-Brand ones, always being asked, can I help you with anything? Are you looking for something? Little do most people know that these individuals have a lack of training in supplements and how they work to really recommend them to anyone. I drew the line when I saw a man, who looked to be 65, walk into another Big Brand store asking for help with energy, drive, and a lack of motivation. He was then recommended an athlete's super vitamin mix and was promised a better life. Seriously? That was for athletes, legitimate athletes! This was potentially going to harm the man's health.

I'm sure I'm not the only one that gets harassed by these questions every time you enter a Nutrition Store. For those that work at these stores, I made this book for you as well, to be a guide for those that want to better their lives. This book is made to be a quick reference for the majority of supplements that can be found being sold in all stores, even including some items that you can find on food labels as well.

I would also like to thank my wife, Amelia. Without her support, there's no way I would have been able to finish this book or dedicate as much time and detailed research as I did. She put up with ridiculous hours and closed-off weekends just to allow me to bring you guys this book, so I cannot thank her enough.

Foreword

This book does not contain any references, nor is it endorsed by any company. The attempt in the writing of this book is to approach it as a nonbiased resource tool. Each individual supplement has been condensed with over 10 resources cross-analyzed, also referred to as a meta-analysis, in an attempt to extract the most commonly seen effects. If I were to include all the resources, it would be larger than this book. Each page has been broken down in the simplest format that is easy to understand by everyone, a format that cannot be found anywhere else.

All information derived from this book comes from various sources, mainly the US National Library of Medicine National Institutes of Health and the National Center for Biotechnology Information further tracked to the source for verification. The information in this book can be found in any medical research website; however, the wording has been simplified and broken down for easier understanding.

I hope this book is of great help in your journey to a healthier and happier you.

Every name is in alphabetical order, with acronyms coming before words written in full. Reference the back of this book to find the Alternative Names Index for names not listed.

Suggested Combinations

Antidiabetics: Zinc, Goldenseal extract (with a high berberine content), Inositol, Cinnamon, and Banaba extract

Anxiolytics (Anxiety reducers): 5-HTP, Ashwagandha, Agmatine, Inositol, GABA, L-Theanine, Agmatine, Rhodiola rosea, and Valerian

Cardiovascular health: Grape Seed Extract, Olive Leaf extract, Pycnogenol, Resveratrol, Garlic, CoQ10 (Ubiquinol/Ubiquinone if over 40) Omega-3 (like fish oil, flaxseed oil or chia seeds), and Vitamin K

Fat loss: Caffeine, Green Tea Extract (with the bioactive EGCG), Synephrine, 5-HTP, and Green coffee extract

Immune Booster: Echinacea, Goldenseal, Garlic, Elderberries, Vitamin C, Zinc, NAC, Astragalus, Spirulina and Noni

Joint Health: Boswellia Serrata, Glucosamine, Chondroitin, Hyaluronic acid, MSM, Collagen (Shark Cartilage), Curcumin (Turmeric extract), Omega-3 (like fish oil, flaxseed oil or chia seeds), Pycnogenol, and Cissus Quadrangularis

Nootropics (Cognitive enhancers): Bacopa monnieri, Caffeine, Guarana, L-Theanine, Creatine, Lecithin (Phosphatidylcholine), Ginkgo Biloba, Green tea extract and Gotu Kola

Mood enhancers: Vitamin D, SAMe, 5-HTP, Ashwagandha, Omega-3 (like fish oil, flaxseed oil or chia seeds), Hordenine, Phenylalanine, St. John's Wort and Rhodiola rosea

Osteoporosis Aid: Magnesium, Vitamin K (preferably MK-4), Vitamin D (with calcium), L-Lysine, Cissus Quadrangularis and Boron (however it may affect hormones)

Performance enhancing: Creatine, Beta-ecdysterone, Beta-alanine, Citrulline, Arginine, Agmatine, Ashwagandha and Caffeine

Testosterone boosters: DHEA, Coleus Forskohlii, Fenugreek, Tribulus terrestris, Maca, D-Aspartic Acid, Vitamin D, Boron, Zinc, and Magnesium

Skin, Hair and Nail Care: Hyaluronic acid, Coconut Oil, Pantothenic acid, Vitamin A, Vitamin K, Astragalus and Zinc

Sleep Aids: Melatonin, Valerian, GABA, L-Theanine (if caffeine was used prior to sleep as it may reduce the stimulatory effects) and Magnesium

Women's Aid: Valerian, Dang Quai, Black Cohosh, Diindolylmethane, Milk Thistle, Flaxseed Oil, Glycyrrhiza, Horny Goat Weed, Inositol (Myo-inositol), Alfalfa and Mexican Wild Yam

Youth Maintenance (50 and over): Vitamin B12, CoQ10 (Ubiquinol/Ubiquinone is preferred), Choline, Phosphatidyl Serine, Lecithin, Lutein, and Goji/Wolf Berries

<u>5-HTP *(5-Hydroxytryptophan)*</u>

5-HTP is a naturally occurring amino acid that's an immediate precursor in the synthesis of the neurotransmitter *serotonin (5-hydroxytryptamine). Serotonin* is an important chemical in the brain that plays a huge part in sleep, mood, appetite, and behavior. 5-HTP supplementation can increase *endogenous (internally created) serotonin* levels by bypassing the *tryptophan hydroxylase* enzyme; this is the rate-limiting/self-regulating step in the creation of serotonin. 5-HTP has been marketed as a mood enhancer and a sleep aid due to its effects on feelings of relaxation. These methods can be extremely effective at raising *serotonin* levels in the body. Common symptoms of low serotonin can be dull muscle aches and pains, depressive episodes, and an increase in weight, as the body tends to over-compensate with carbohydrate-rich diets, to include sugary foods. (Although many of these symptoms can be from another underlying factor that should be discussed with a medical professional.)

In the body, 5-HTP is able to cross the blood-brain barrier and is also converted into *Serotonin* through the enzyme *L-aromatic amino acid decarboxylase (AAAD);* it should be noted that *dopamine* is ALSO synthesized by the same enzyme which could block *dopamine* synthesis. (Read the note at the bottom).

- Additionally, *dopamine* and *serotonin* ALSO share the same metabolic breakdown pathway, monoamine oxidase; sharing this enzyme does not establish that raising *serotonin necessarily dep*letes *dopamine* in humans.

With that note, it is also important to mention that 5-HTP can have interactions with **L-DOPA** as it also shares *AAAD* which can lead to depletion of each, *dopamine* and *serotonin*; **L-Tyrosine is a precursor to L-DOPA, but combining precursors is not** an established way to balance neurotransmitters.

<u>Lastly, it has been proposed that long-term 5-HTP supplementation could alter *catecholamine (dopamine, adrenaline, and noradrenaline)* levels in the body; this is not an establish*ed* explanation for severe depression or a guide to supplement combinations. 5-HTP supplementation for the treatment of depression should be done under medical supervision.</u>

<u>Proposed benefits (evidence varies)</u>

- Anxiolytic (anti-anxiety)
- Improved mood
- Antidepressant
- Decreased appetite
- Possible sleep aid

<u>Possible Side Effects</u>

- Heartburn
- Diarrhea
- Nightmares
- Nausea
- Drowsiness
- Vomiting
- Sexual Problems

<u>Dosage</u>

- Studies use varied doses; 300–500 mg/day is not a general safe or effective recommendation.
- No established effective weight-loss regimen exists; depression treatment needs medical supervision.

<u>***WARNINGS***</u>

Seek medical advice before use. 5-HTP may affect those who are on MAOI or SSRI antidepressants; please consult a doctor prior to use. Combining serotonergic medicines with 5-HTP can cause serotonin syndrome, including at usual doses; avoid this combination without specialist direction.

7-Keto-DHEA *(7-Ketodehydroepiandrosterone)*

7-Keto-DHEA, also known as 7-oxoprasterone, is a non-androgenic metabolite of **dehydroepiandrosterone (DHEA)**. Although it is not directly converted to testosterone, it is not established as more effective or safer than DHEA. 7-Keto-DHEA is naturally occurring in the body, but levels vary greatly from person to person. It can be seen marketed as a liquid or in a powdered form like capsules.

7-Keto-DHEA is not an established anabolic *prohormone* and does not directly convert into testosterone or estrogen. The conversion from DHEA to 7-Keto-DHEA by the enzyme *cortisone reductase* is irreversible as this pathway is not multi-directional. This means that it cannot be turned back into DHEA, and since DHEA is the intermediate and basis for androgenic and estrogenic hormones, 7-Keto-DHEA cannot be further converted into androgens or estrogens.

- 7-Keto-DHEA may affect cortisol levels, as mentioned before; it uses the enzyme *cortisol reductase,* which is involved in the conversion of *cortisol (the stress hormone)* and *cortisone (a cortisol metabolite).* The enzyme is multi-directional, which can create *cortisol* from *cortisone*; 7-Keto-DHEA can be a competitive selector of this enzyme, which may ultimately lead to a reduction of *cortisol. Theoretically,* this can increase testosterone to cortisol ratios; concurrently, this is where the enhanced muscle promoting environment effects that are advertised come from. While it may directly not be an androgenic compound itself, it may indirectly increase testosterone levels, although there currently is a lack of clinical data to support this.

Small short trials suggest possible weight effects, but sustained loss and long-term safety remain uncertain. Proposed thermogenesis involves effects on enzymes involved in *mitochondrial* activity and efficiency. The activation of these enzymes can actually reduce *mitochondrial* efficiency, particularly the *Krebs Cycle (This is the chemical process in which energy is released through oxidation creating ATP).* This, in turn, can cause a demand for more fatty acids or glucose to be oxidized to reach near-normal output levels of **ATP** for use in the body. (This is a proposed efficiency change, not a measured fatty-acid-to-ATP conversion ratio.)

Proposed benefits (evidence varies)

- Promotes muscle growth
- Promotes lean muscle
- Improves fat loss
- Increased calorie burn

Possible Side Effects

- Acne
- Oily skin
- Perspiration
- Hair thinning or hair loss
- Insomnia
- Headache

Dosage

- 200-400 mg a day for 4 weeks, dosing twice a day

WARNING

Seek medical advice before use. 7-Keto-DHEA may affect those who are diagnosed with **Endocrine Abnormalities** as 7-Keto-DHEA may affect hormone levels. It has also been seen to alter **thyroid hormone** levels and **adrenal** function.

AAKG *(Arginine Alpha-ketoglutarate)*

AAKG is basically Arginine attached to the ketone alpha-ketoglutaric acid, due to their importance in vital body functions. Arginine is a conditionally essential amino acid that is involved in nitric oxide levels while alpha-ketoglutaric acid plays a role in the *Krebs Cycle.* However, outside of the individual molecules' biological activities, evidence for the combined substance's purported athletic-enhancing effects is lacking; it has failed to demonstrate any enhancement in clinical trials.

In the theoretical concept of this supplement, it would make sense that it could potentially promote these effects. Alpha-ketoglutaric acid and arginine are two main components involved in the metabolism of nitric oxide, with alpha-ketoglutaric acid also playing a role in the creation of **ATP** in the *Krebs Cycle.*

- With the addition of alpha-ketoglutarate, it is believed that performance can be improved via increasing the rate of **ATP** synthesis through enhancing the *Krebs Cycle.* This may be done through the addition of alpha-ketoglutarate, as it is an intermediate in the cycle that oxidizes *acetyl-CoA (derived from fats, carbohydrates and proteins),* supporting ATP production. Additionally, the conversion of alpha-ketoglutarate to *succinyl-CoA* by the enzyme *alpha-ketoglutarate dehydrogenase* creates *NADH, which* can be further oxidized for ATP use. Theoretically, this should increase and improve energy levels in the body. If Alpha-ketoglutarate levels are deficient for the *Krebs Cycle,* it can be replenished from **glutamine** found in muscle tissue as glutamine is made up of alpha-ketoglutarate.
    - Glutamine is abundant in skeletal muscle's free amino acid pool, not 60% of muscle tissue. It transports *nitrogen* and supports protein metabolism.
- In the *Urea Cycle,* arginase converts arginine to **ornithine** and releases urea. Ornithine combines with carbamoyl phosphate to form **citrulline; citrullin**e then returns to arginine through argininosuccinate. The cycle disposes of nitrogen from ammonia and aspartate as *urea. Urea* is far less toxic than ammonia. This pathway is relevant to nitrogen handling, but does not establish that AAKG supplements improve energy or endurance.

As of now, human exercise results are mixed, with no consistent established improvement in muscular performance.

Proposed benefits (evidence varies)

- Vasodilator
- Enhance endurance
- Increase strength

Possible Side Effects

- Gastrointestinal discomfort
- Low potassium levels
- Palpitations
- Increased stomach acid
- Headaches
- Nausea
- *Syncope (temp. blackout)*
- Vomiting

Dosage

- The recommended dosage is 3.5 to 6 grams.

WARNING

Seek medical advice before use. AAKG may affect those who are diagnosed with **low blood pressure** as AAKG may lower pressure to unsafe levels; AAKG may also affect those going through **herpes** treatments.

<u>ALA (Alpha-Lipoic Acid)</u>

Alpha-Lipoic Acid, not to be confused with the **omega-3** fatty acid *Alpha-Linolenic Acid,* is an organosulfur (contains sulfur) compound. ALA is normally synthesized by enzymes in the *mitochondrion* from *octanoic* acids in animals and is a necessary *mitochondrial* compound as it plays a role as a cofactor for *alpha-ketoglutarate dehydrogenase (reference the previous page)* in the *Krebs Cycle* (**ATP** creation). Currently, it has been marketed as an antioxidant, fat-loss aid and a possible antidiabetic. A 2024 Cochrane review found little or no neuropathy benefit after at least six months; route and **ou**tcome matter.

- <u>R-ALA is the naturally occurring stereoisomer used by mitochondrial enzymes. Many synthetic products contain both R- and S-ALA. S-ALA is not simply proven to cancel half the benefit, and greater R-form bioavailability does not establish superior clinical outcomes for every use. R+ ALA or RLA refers to the R form; S-ALA refers to the other stereoisomer.</u>

ALA acts as a potent antioxidant as it is able to protect the body from oxidant damage as well as a metal chelator. Its antioxidant activity has been seen to be derived from a couple of ways:

- It is able to increase *mitochondrial* oxygen capacity, meaning it is able to increase the capacity of oxygen that is used in energy production. This can potentially lower oxidation from the lack of oxygen free radicals produced due to effective oxygen use in energy production. Additionally, it has been seen to increase **glutathione** levels through increasing levels of the rate-limiting enzyme *glutamyl cysteine ligase*; glutathione is an *endogenous (inside the body)* antioxidant.

ALA has been seen to act as a chelator along with its metabolite *dihydrolipoic acid*; they have shown keen attraction to bind to the following metals: lead, zinc, copper, **iron**, mercury, **selenium,** and **manganese**. ALA toxicity is not reliably identified or treated by assuming selenium or manganese deficiency; suspected toxicity requires medical assessment.

Outside of a powerful health product, ALA has also been seen and associated as a food partitioning agent, meaning it is able to increase the distribution of nutrients into muscle tissue instead of fat storage. Particularly glucose, ALA has shown enhanced glucose uptake into muscle cells by recruiting more *GLUT-4 (glucose transporter type 4)* receptors from the *Golgi* to the membrane of striated muscle cells.

<u>Proposed benefits (evidence varies)</u>

- Antioxidant/metal chelator
- Food partitioner
- Antidiabetic
- Promote muscle growth
- Anti-inflammatory
- Aid fat loss

<u>Possible Side Effects</u>

- Nausea
- Skin rash
- Dizziness

<u>Dosage</u>

- 50-200 mg once a day for antioxidant effects
- 67-400 mg 3x day for food partitioning/insulin sensitivity prior to each meal

<u>***WARNING***</u>

Seek medical advice before use. ALA may affect **Diabetics** as ALA can lower blood sugar levels; medication may need to be modified. ALA might interfere with **Thyroid** treatments. ALA has been known to lower thiamine levels in the body; consumption of alcohol while taking ALA may dangerously lower thiamine levels in the body.

ATP *(Adenosine 5'-Triphosphate)*

ATP is an organic chemical and the primary source of systemic energy in the body; it has been marketed as a performance enhancer although supplementation does not seem to affect ATP levels in the body. Any effect that is accomplished may be through increasing uric acid in the body, which may very well act just like an antioxidant; benefits have been seen as insignificant at best. (Read the warning note.)

Regardless, here's a brief explanation of how ATP functions within the body. ATP is a molecule that works as a three-step biochemical decomposition, delivering energy where energy is needed in your body — usually by hydrolysis of ATP to Adenosine Diphosphate and inorganic phosphate; further reactions can form Adenosine Monophosphate. Your muscle tissues need energy to perform. The more your muscles contract and stretch the more ATP is burned up. ATP comes from different sources of your body depending on how you train.

Phosphagen System

- Through this system, ATP is used rather quickly, 3 to 10 seconds quick. When this is activated, your muscle tissues have quick access to ATP. **Creatine** Phosphate is used to replenish ATP reserves, but as stated above this only continues for 3 to 10 seconds.

Glycogen-Lactic Acid System AKA Anaerobic System

- This is the second stage of energy, Anaerobic meaning "no oxygen." In short, you use sugars/carbohydrates known as glycogen. Glycogen is then converted to glucose. This is a rather complex process, so it takes a little longer to replenish ATP at this stage. Lactate regenerates NAD+ so glycolysis can continue. It is not the cause of delayed muscle soreness, and foam rolling does not work by clearing trapped lactic acid. This energy conversion lasts anywhere from 60 seconds to 2 minutes.

Aerobic Respiration

- All energy systems work together; aerobic contribution rises with duration rather than switching on at two minutes. At this point, oxygen breaks down glucose into two separate compounds, carbon dioxide and water. Fat and carbohydrate oxidation occur together in proportions set by intensity, duration and fuel availability; fat use does not require complete glycogen depletion. Protein usually contributes less.

Proposed benefits (evidence varies)

- Improved muscle growth
- Improved stamina
- Improved strength

Possible Side Effects

- Headache
- Low blood pressure
- Chest pain
- Sweating
- Nausea
- Breathing problems
- Flushing
- Anxiety
- Sleep problems

Dosage

Average dosages seen in supplements; regardless, no effective dosage has been seen

- 125-250 mg 2x a day

WARNING

Seek medical advice before use. ATP may affect people with **Gout**, **kidney stones** or **heart disease**, as ATP can increase uric acid in the body which can increase crystal deposits in tissues increasing the risk and severity of these conditions.

Acai *(Euterpe oleracea extract)*

Acai, whether referring to the berry, the extract, or the plant, Acai comes from the Acai Palm found in the *Arecaceae* family. The berries are small, darkish purple and round generally around the size of a marble. Just to clear a quick confusion, the correct spelling can be seen as *açaí*, but it will be referred to as acai from here on forth just to simplify it. A couple of years ago acai berries and extracts were on the craze as a "maximum antioxidant" profile, claiming reduced fat accumulation and various other bold statements. Although the berry itself does have various health benefits, it is no more effective than the blueberry that you could find at your local grocery store. Fruits generally have various antioxidant profiles, some being more beneficial in particular areas, but none are a miracle superfood.

The pulp in the berry contains various minerals and vitamins, along with some *anthocyanins* and *flavonoids*, the antioxidant portion of the berry.

- The *flavonoid*, particularly *velutin*, has shown anti-inflammatory activity in laboratory studies by inhibiting *interleukin-6 (IL-6)* and *Tumor Necrosis Factor-Alpha (TNF-α)* inflammatory *cytokines (inflammatory messengers)* found on white blood cells, specifically *macrophages*.
- The *flavonoid apigenin has* proposed *neuroprotective actions in preclini*cal research; brain entry alone does not establish human benefit. It has been seen to reduce *excitotoxicity* in neurons by regulating calcium ion (an increase of calcium ions causes excitability in neurons) influxes in neural *mitochondria* reducing the creation of *Reactive Oxygen Species (an oxidant)* by dysfunctional *mitochondria*.
- There is a belief that acai berry extract has systemic anti-inflammatory properties, and though it has been seen to interact and downregulate *cyclooxygenase-2, IL-6, IL-1β* and *TNF-α (all of which are pro-inflammatory messengers),* various studies have been inconclusive, with some showing support for this effect while others state the effects are insignificant. Antioxidant activity has been known to reduce oxidant-induced inflammation.

Through the mechanism mentioned above, the belief that Acai berry is a powerful anti-fat supplement emerged; enhanced mitochondria can, in theory, increase caloric consumption and glucose utilization which could ultimately lead to fat loss. However, fat loss has not been directly linked to Acai berry or extract consumption, just proper mitochondrial function.

Proposed benefits (evidence varies)

- Anti-inflammatory
- Antioxidant
- Neuroprotection

Possible Side Effects

- Nausea
- Gastrointestinal discomfort

Dosage

- 1,000 mg capsules once or twice a day or Acai berry serving size is 3.5 ounces (100 grams)

WARNING

Seek medical advice before use. Acai Berry may affect **Diabetics** as Acai may affect blood glucose levels as it can interact with glucose, medications may need to be modified. Raw Acai berry juice may contain a parasite that can cause **American trypanosomiasis** or **Chagas Disease; symptoms** are similar to that of the flu.

Acetyl L-Carnitine (ALCAR)

ALCAR is an *acetylated* form of **L-carnitine** and is an amino acid-derived compound, rather than a mixture of a vitamin and an amino acid synthesized in the body through **lysine** and **methionine**, making it essential for body functions. The body can convert L-carnitine and **Acetyl-L-Carnitine (ALCAR)** into various derivatives dependent on need. ALCAR can be seen sold and marketed as a fat loss agent, antioxidant, and performance enhancer. It has additionally shown neuroprotective promise. (Reference L-Carnitine for further L-Carnitine benefits as they still apply to ALCAR; this page applies to ALCAR **only**.)

Acetylated carnitine is able to cross the blood-brain barrier; ALCAR has been seen to increase glucose and creatine usage in the brain resulting in increased cognitive energy levels. ALCAR has also been seen to provide neurological protection by prolonging the antioxidant effects of the enzyme *Superoxide Dismutase* with the addition of promoting the synthesis of the neurotransmitter *acetylcholine (responsible for motor function and brain function)* due to the acetyl group attached to ALCAR.

- The mechanisms described above seem to be accomplished through ALCAR metabolization and dissociation of carnitine and its acetyl group through the *Pyruvate Pathway.* This can enhance the intermediates involved in the *Krebs Cycle,* as the *Pyruvate Pathway* creates various metabolites for the *Krebs cycle.*

The *Krebs Cycle* is used to release stored energy through the oxidation of *acetyl-CoA; CoA* needs to become *acetylated* prior to its oxidation. This is where ALCAR is able to have synergistic benefits with the mechanism above since it is able to donate its acetyl group. Once this is accomplished, *acetyl-CoA* is oxidized and broken down to release energy which is used for **ATP** in the *mitochondria.*

ALCAR can also be seen as a performance enhancer as it is able to reduce *mitochondrial* decay and promote *mitochondrial* function. Carnitine can enhance mitochondrial function as it facilitates the transport of substrates for fuel, such as fatty acids, *adipose* (fat) tissue, and glucose to the mitochondria, and in the exact same mechanism as mentioned above it is able to use its acetyl group for use as energy.

Proposed benefits (evidence varies)

- Performance enhancer
- Vasodilator
- Fat Loss
- Neuroprotector
- Improved recovery
- Mental Clarity

Possible Side Effects

- Gastrointestinal discomfort
- Nausea
- Vomiting
- Headaches
- Abdominal cramps
- Trouble sleeping
- **Psychosis (those diagnosed with bipolar disorder)**

Dosage

- Studied oral ALCAR amounts commonly range from 500 to 3,000 mg/day; the effective regimen depends on the condition and formulation.
    - Tends to be more effective with a meal, 30-40 grams of both carbohydrate and protein

WARNING

Seek medical advice before use. ALCAR may affect **Diabetics** as ALCAR can affect blood sugar levels; medications may need to be modified. If you have **Blood pressure** problems, ALCAR can lower blood pressure. ALCAR can also interact with **Beta-Blockers** or **Calcium-channel blockers** as it can affect epinephrine/norepinephrine.

Acidophilus *(Lactobacillus Acidophilus)*

Interestingly enough, all the information that you need is the name. The exact translation of this word is an acid-loving rod-shaped bacterium that lives in milk, perfectly enough a probiotic. These are gram-positive bacteria that can be found naturally occurring in the digestive tract and mouth of animals and humans, as well as in the vaginal microbiota of female sexual organs. Commercially, it has been used for various dairy products, with the most prominently known being yogurt. It plays a huge role in the human body in converting sugars to lactic acid.

Acidophilus and its interaction with the body have been seen to provide various benefits as it can promote *mucin (glycoproteins that create gel-like secretions in the intestines)* secretion. This is important as it helps with cell signaling and provides a chemical barrier for *epithelial (lining of tissues)* tissues. In particular, Acidophilus can promote the secretion of the *Chloride Anion Exchanger (membrane protein in intestinal cells involved in electrolyte transfer)* protein; this enhances intestinal epithelial barrier function and can promote electrolyte absorption in the small intestine, especially in cases where dehydration may occur like diarrhea.

- It is believed that Acidophilus can express a *substrate (substance)* that may be able to promote *ion (pertaining to electrolytes and minerals)* exchange through a related Phosphatidyl *3-kinase-related kinase (involved in various molecular/genetic functions)* mechanism.

These little bacteria help in the absorption of nutrients and to keep all the bad bacteria and yeast at bay.

Through its mutual existence in humans, it puts less stress on the liver and kidneys through anti-inflammatory and antioxidant capabilities. It seems to have potential as an antioxidant capable compound; it has been seen that Acidophilus can increase levels of the antioxidant **Glutathione**, *Glutathione peroxidase*, and *Super Oxide Dismutase* enzyme, all of which reduce oxidative and peroxide damage to cells and tissues in the body. This has led to a study as a possible anti-arthritic. However, more needs to be assessed to see the potential as an anti-arthritic.

Proposed benefits (evidence varies)

- Improve nutrient absorption
- Promote digestion
- Anti-diarrheal
- Possible anti-inflammatory
- Possible antioxidant

Possible Side effects

- Gas
- Bloating
- Diarrhea

Dosage

- There is no universal one-billion-organism ceiling; select a studied strain, indication and dose. If in capsule form, the capsule should have a coating on it; this helps bacteria enter the intestines unharmed.

WARNING

Seek medical advice before use. Acidophilus may affect people with **Immune deficiencies or stomach disorders** as Acidophilus is a gram-positive bacterium and could affect weakened immune systems and cause complications with stomach disorders. Serious illness or impaired immunity increases concern about rare invasive infection. Benefits and risks are strain- and patient-specific; no universal 10^9-cell limit is established.

African Mango *(Irvingia gabonensis or IGOB131)*

The African Mango, as it's referred to, is not a mango at all but seeds inside the fruit of the African Mango tree from the *Irvingiaceae* family. Although they do bear mango-like fruits that are rich in fat and protein, the supplement you are buying at the store is actually the extract of the tree. The fruit can be processed into jellies and jams, as well as wines and color dyes and eaten just as any other fruit and similar to other fruits, it provides generic benefits and does not really possess any medicinal or highly beneficial properties. The extract has mainly been advertised as a fat loss product, although most of the products do not contain real African Mango extract and the real extract itself is not very potent at inducing fat loss.

The main bioactive compounds found are *ellagic acid* and *ellagitannins; it* is important to note that there are many variations of this plant and its extracts; it seems most supplement manufacturers have not come to a conclusion on which part of the tree provides the most benefit in terms of an extract. With that said, this is a description of the constituents found in the seeds and therefore the extract derived from the seeds.

- *Ellagic acid* and *ellagitannins* are natural polyphenols that can offer antioxidant activities that have free radical scavenging activities, although they are broken down into *urolithins,* the metabolite itself has fairly weak antioxidant capabilities. As mentioned previously, these are not specific to this plant. The antioxidants found in this plant can be found in other fruits.

The actual supplements out there, however, are not that. This particular extract is fairly new and comes from the seeds. The few short studies suggest possible fat loss, but study quality limits confidence in obese or otherwise heavier-set persons without much change in lifestyle. Like mentioned before, the majority of supplements out there do not have a set extract; some are derived from the bark of the tree while others from the peels, and the only ones that have had a molecular examination are the extracts derived from seeds which have been seen to offer insignificant value to the user.

Proposed benefits (evidence varies)

- May reduce LDL cholesterol
- May stop fat production
- May promote fat loss

Possible Side Effects

- Headache
- Dry mouth
- Sleep disturbance
- Gastrointestinal discomfort

Dosage

- As of current, there is no set dosage; amounts taken have ranged anywhere from 150-3,000 mg

WARNING

Seek medical advice before use. African Mango may affect **Diabetics** as African Mango may affect blood sugar levels; medication may need to be modified. Those on **hormone therapy** or **hormone-sensitive conditions** like **endometriosis**, **ovarian**, **uterine,** or **breast cancer** should exercise caution as African Mango has been seen to interact with hormones. For those taking **hormone boosters**, African Mango may negatively affect testosterone synthesis; supplements include:

- Fenugreek (Trigonella foenum-graecum)
- Guar Gum
- Siberian Ginseng
- Devil's Claw
- Asian Ginseng (Panax ginseng)

Agmatine *(4-aminobutyl guanidine)*

Agmatine is a substance that is naturally created from **Arginine**. Agmatine can be found in various pre-workouts and has been marketed mainly for its vasodilation effects, amongst other less common benefits: a fat loss aid, an *analgesic (pain killer)* and a possible food partitioner. It seems to exert its effects mainly as a neuromodulator.

Agmatine has been seen as a potent *analgesic* which can aid in various situations mainly touted to aid difficult training sessions as a buffer of fatigue, which may be the reason as to why it has recently been looked upon as an alternative to pain relievers. Its mechanism seems to be derived from its ability to promote *β-Endorphin* release, as *β-Endorphins are opioid receptor agonists; β-Endorphins can attach themselves* to the *opioid receptors* of cells in the area where pain was perceived, reducing the pain signal; this can provide a global relief/function resulting in reduced pain throughout the body.

- Since Agmatine has been observed as a *non-selective ligand (binds/activates the receptor),* an additional benefit seen is that it has been observed as a possible activator of *imidazoline I2 receptor* and possibly *imidazoline receptor I3* that regulates insulin secretion in *beta cells.* This may be the basis for its partitioning effect on blood glucose along with interaction with *β-Endorphins* which may also add benefit and regulate glucose uptake into muscle tissue. Currently, however, its actual mechanism isn't currently known.
- Agmatine seems to also offer antidepressant and anti-anxiolytic benefits as it is able to induce proper *locus coeruleus (a cluster of neurons in the upper/center of the brain stem involved in emotions)* neuron firing at moderate doses with the addition of the modulation of *β-Endorphins, it* is able to have antidepressant/anxiety effects as well as a synergistic effect with antidepressant drugs.

Agmatine also stimulates nitric oxide synthesis through *α2A receptor* activation, causing a cascade signaling which ultimately calls for the creation of Nitric Oxide synthesis and its sustainment as well.

Benefits

- Enhance performance
- Improve endurance
- Vasodilator
- Aid fat loss
- Nootropic
- Anxiolytic

Possible Side Effects

- Gastrointestinal discomfort
- Vomiting
- Diarrhea

Dosage

- For nootropic (mental) effects, users reported a recommended dosage is 500-750 mg
- For pain tolerance users reported 600- 750 mg
- For rigorous training users reported 500 mg to 2.5 grams

WARNING

Please consult with a medical professional prior to administering Agmatine. There is concern that Agmatine may affect **Diabetics** as Agmatine has been shown to affect blood sugar levels; medication may need to be modified. Agmatine may also increase **SSRI** effects; medication may need to be modified. Agmatine can also affect those with **blood pressure problems** as it has been known to lower blood pressure.

Alfalfa *(Medicago Sativa)*

Alfalfa, also known by the name Lucerne, is part of the *Fabaceae* family where legumes like peas, peanuts and lentils can be found. It is native to South Central Asia and Iran. It averages a height of around 3 feet and grows purple blue-ish like flowers that spiral about 2 to 3 turns when they reach maturity. Alfalfa has been used as a universal crop, from hay to feed livestock as far as being used to improve soil composition. It's also an excellent source of protein, minerals, and vitamins, to name a few:

Nutrition Content	1 cup (33g) of Alfalfa	Nutrition Content	1 cup (33g) of Alfalfa
Protein	- 1.3 g	Calcium	- 10.6 mg
Vitamin A	- 51.1 IU	Potassium	- 26.1 mg
Vitamin C	- 2.7 mg	Phosphorus	- 23.1 mg
Vitamin K	- 10.1 mcg	Zinc	- 0.3 mg
Folate	- 11.9 mcg	Magnesium	- 8.9 mg
Choline	- 4.8 mg	Manganese	- 0.1 mg
Omega-3	- 57.8 mg	Omega-6	- 77.2 mg

Additionally, it contains the enzymes *amylase (breaks down starches), pectinase (breaks down pectin)* and *invertase (breaks down sucrose)* that can aid digestion.

It was also used in traditional Ayurvedic medicine to improve cell *proliferation (multiplication),* although currently it has been seen to only affect white blood cells, *neutrophils, lymphocytes* and *granulocytes* through an increase of *Superoxide Dismutase (antioxidant).* It is believed that Alfalfa can also improve insulin sensitivity in Diabetics mainly through its high manganese content as manganese has been shown to improve *Beta cell (Regulates insulin secretion)* towards baseline/normal function.

Alfalfa has also sometimes been used to treat menopause symptoms, as it contains the *phytoestrogens genistein, daidzein* and *coumestrol* which have been seen to bind, although weakly, to estrogen receptors; this weak stimulation may provide relief to menopausal or menstrual symptoms only.

Proposed benefits (evidence varies)

- Menopause aid
- Digestive Aid
- Antidiabetic

Side Effects

- Contact dermatitis

Dosages

- Dry leaves are 500 to 1000 mg or 1 to 2 ml of standardized extract a day.

WARNING

Seek medical advice before use. Avoid in p**regnan**cy: stachydrine and homostachydrine are alkaloids with suspected uterine effects. Seed L-**canavanine** has been linked to lupus-like symptoms and reduced blood-cell count**s**. Avoid with lupus; Vitamin K can interfere with warfarin. Review glucose-lowering medicines and phytoestrogen-sensitive conditions, including breast, uterine or ovarian cancer, endometriosis and adenomyosis.

Apple Cider Vinegar (ACV)

ACV is made from the fermentation of apple juice into alcohol then from alcohol to vinegar through the use of *Acetobacter* bacteria with a very low acidic value of 5% *acetic acid.* Its various touted health values come from its vinegar and its antioxidant profile as it does not contain much nutritional content outside of that. It's important to know the difference between Raw ACV and the typical ACV you would buy at a grocery store. How do you tell the difference? One is raw and unpasteurized while the other is not. You can instantly tell by looking at it. Is it cloudy? Does it have a web-like look to it? This is normal, and that's what you're looking for, and it actually makes the taste a bit sweeter; cloudiness does not establish greater health benefit than filtered vinegar.

Through use, it has been shown to increase insulin sensitivity if used prior to a carbohydrate-rich meal, mainly because it can slow the absorption of glucose; this may ultimately lead to improved insulin sensitivity. It does not abolish carbohydrate absorption; it just slows it, as with any vinegar, it slows the stomach emptying rate. Small meal studies do not establish durable diabetes control or reliable weight loss, and vinegar must not replace treatment.

In laboratory experiments, ACV has been seen to amplify immune response and enhance *monocyte (type of white blood cell)* response to microbes, while reducing *proliferation (multiplication)* of various microbes as it is capable of starving and affecting gene regulation as well as intracellular metabolic mechanisms as it is able to break down the ribosomes *(synthesize proteins)* of some microbes.

- Along with its antioxidant content of *flavonoids, caffeic acid, gallic acid,* and some others, it can reduce oxidative and pro-inflammatory responses brought on by microbes by down-regulating *Il-6* and *TNF* pro-inflammatory *cytokines (cell messengers)* while increasing *phagocytosis (devouring pathogens)* in *monocytes.*

<u>Proposed benefits (evidence varies)</u>

- Antidiabetic
- Anti-microbial
- Fat loss
- Immune booster
- Digestive aid

<u>Side Effects</u>

- May reduce potassium levels
- Throat/mouth irritation
- An increase in tooth sensitivity due to the acidity
- Slow digestion
- Tooth sensitivity
- Nausea
- Acid Reflux

<u>Dosage</u>

- 1 tablespoon per 1 cup of water prior to a meal
  - It's not recommended to take apple cider vinegar by itself, as it can reduce tooth enamel and irritate the soft tissues in your throat and mouth.

WARNING

Seek medical advice before use. Apple Cider Vinegar may affect **Diabetics** as Apple Cider Vinegar has been seen to reduce glucose levels and affect insulin sensitivity; medication may need to be modified. Large doses of Apple Cider Vinegar may affect those with the risk of **Osteoporosis** as excessive intake has been associated with low blood potassium; this is not loss of potassium specifically from bones.

Apple Pectin

Apple Pectin, or any pectin for that matter, is a semi-soluble indigestible fiber derived from fruits like citrus and apples, with apples generally having more pectin than any other fruit. Pectin has been universally used in the process to make jelly-like products, such as jam or medical preparations, and as a stabilizer in various food items. Currently, it has been used and marketed as a gastrointestinal health agent as it can regulate intestinal health.

The fiber is considered a *polysaccharide (carbohydrates)* that is derived from the cells' walls, as they are tough, flexible, and rigid, in plants. When consumed or when water is added, the Pectin becomes gel-like, and for humans, it cannot be digested. As such, it's good at helping maintain a healthy digestive system as it has the capability of cleaning and maintaining proper health for intestines.

The gel-like substance can inhibit the absorption of fats and carbohydrates when consumed along with a high-fat content meal while additionally increasing levels of *satiety (feeling full)* from its gel-like content. Additionally, it is able to maintain bacterial balances in the gut by maintaining proper levels of the bacteria *Bacteroides (ferments carb-rich food into absorbable fatty acids)* and *Firmicutes (involved in energy resorption and breakdown of insoluble fibers)*, both of which make up the majority of the intestinal flora.

Apple Pectin consumption has been seen to reduce diet-induced intestinal inflammation, particularly the *ileum (the end of the small intestine)*, through inhibiting the inflammatory signaling derived from *Toll-Like Receptor 4 (TLR4)* activation. This is a protein that signals for inflammatory *cytokine (cell messengers)* production.

- It may accomplish this by promoting protein levels involved in intestinal permeability integrity, thereby reducing diet-induced *metabolic endotoxemia*. This is where "bad bacteria" are able to cross the intestinal barrier thereby causing system-wide inflammatory and oxidative response through the mechanism mentioned above, as the breakdown of bacteria, like gram-negative bacteria, releases *Lipopolysaccharide (LPS)* an endotoxin that activates TLR4.

Benefits

- May lower cholesterol
- Reduce gallstones risks
- Support intestinal health
- Regulate blood pressure
- Antidiabetic

Possible Side Effects

- Gas
- Diarrhea
- Cramps

Dosage

- Dosage varies by reason for use; 5-15 grams daily can support general body/intestinal health

WARNING

Please consult with a medical professional prior to administering Apple Pectin. There is concern that Apple Pectin may affect those with **Asthma** as inhaling Pectin could worsen symptoms of asthma. Apple Pectin may also affect **diabetics,** as it may reduce the requirement of insulin levels; medication may need to be modified. Because apple pectin slows the enzymes that break down starches and sugars, this could give you unusual sugar readings.

<u>Arachidonic Acid *(cis-5,8,11,14-eicosatetraenoic acid)*</u>

Arachidonic Acid (AA) is an omega-6 *Polyunsaturated Fatty Acid (PUFA)* that can be found in phospholipids, a component of cell membranes, and is a derivative of *Linoleic Acid*. AA plays a role in inflammatory signaling and vasodilation. The body is capable of recognizing training-induced inflammation from inflammation caused by injuries. Acute inflammatory responses that arise from training can be extremely beneficial for growth as it can promote blood flow to the affected muscle optimizing nutrient delivery.

AA can act in either one of three ways: the *Lipoxygenase (enzymes that catalyze the oxidation of PUFA), Cyclooxygenase (enzymes that make inflammatory, clotting and vasodilating signals),* or *P450 (can metabolize PUFAs into active cell signaling molecules)* pathways. When you lift or train in any shape or form that involves rigorous activity, your muscles become damaged. That's when enzymes flock over to the damaged area and release AA in the affected area; AA is then converted into *prostaglandin* by *Cyclooxygenase 1 & 2* enzymes. *Prostaglandin* is responsible for the pain and inflammation you feel the day after; this is the process the body uses to increase localized nutrient delivery to repair damaged muscle tissue, boosting protein synthesis. Training is needed to achieve enhanced muscle effect.

- When NSAIDs (Anti-inflammatories) are consumed, this entire process becomes interrupted, which, in response, might cause a "plateau effect."

It is important to note that the other two pathways can still be activated; activation of the *P450* Pathway creates *epoxyeicosatrienoic acids* and *hydroxyeicosatetraenoic acids*. Both of the metabolites affect blood pressure which can affect those with a family history of cardiovascular problems.

- They can also promote vasodilation as *epoxyeicosatrienoic acid* is a vasodilator.

Lipoxygenase pathways increase inflammatory signals which can result in aches and pains in joints that may have been unnoticeable before. <u>A combination of all three mechanisms could affect dormant family health histories such as *demyelination* of nerves and neurons; additionally, high amounts of AA can become cytotoxic to neurons and nerve cells.</u>

<u>Benefits</u>

- Promote muscle growth
- Reverse plateaus
- Increase strength
- Vasodilator
- Increase in power

<u>Possible Side Effects</u>

- Low/high blood pressure
- Mood swings
- Inflamed/red spots
- Poor concentration
- Joint pain
- Frequent urination

<u>Dosage</u>

- To improve athletic performance 1.5 g a day is recommended
- Omega-3 fatty acids can enhance AA properties while maintaining a proper omega ratio.

<u>***WARNING***</u>

Please consult with a medical professional prior to administering Arachidonic Acid. There is concern that Arachidonic Acid may affect those with pre-existing inflammatory conditions such as **Arthritis**, **eczema**, **breast tenderness** as it can augment inflammatory signals. **For those with Diabetes or Kidney problems, AA** can put additional stress on kidney function, which may worsen the condition.

Artichoke Extract *(Cynara scolymus)*

Artichoke *Cynara scolymus* is the edible artichoke that is commonly consumed in the West, e.g., United States and is native to the Mediterranean. The plant's extract comes mainly from the plant's large lobed leaves, as this part of the plant has the highest amounts of bioactive compounds and has been used in traditional medicine for indigestion. Currently, it has been seen marketed as a weight loss and *satiety (feeling of fullness)* product, although, its effects seem minimal at best, which is why it would be seen combined with **White Kidney Bean** extract as they seem to have synergistic effects.

The main bioactive compounds seen in this plant are *luteolin* and *caffeoylquinic acids.* The extract seems to inhibit the *HMG-CoA reductase enzyme. HMG CoA reductase* is an enzyme that produces cholesterol; inhibition of this enzyme increases cholesterol receptors in the liver; this results in the breakdown of cholesterol and lower levels of cholesterol in the body. This effect, while consistent, is mild.

- *Luteolin* seems to offer additional cardioprotection through exerting antioxidant effects by protecting the *endothelium (lining of tissues)* and *monocytes (a type of white blood cell)* by reducing the *Low-Density Lipoprotein (LDL)* oxidation and by preventing their oxidation/*foaming.*
    - *LDL can attach to white/red* blood cells causing them to become *foam cells. Foam cells* can then start to collect onto blood cells as they are extremely sticky, which can lead to *atherosclerotic* build-up. (Or plaque build-up in the arteries for short.)

By decreasing oxidation, the risk of *foam cells* decreases as well, thereby decreasing the risk of cardiovascular disease.

- Artichoke extract has also been seen to reduce *Vascular Cell Adhesion Molecule-1* and *Intercellular Adhesion Molecule-1*, both of which are involved in the build-up and adhesion on the lining of vascular and non-vascular tissue; this can result in enhanced blood flow.
- *Nitric Oxide* boosting properties come from the downregulation of *Nitric oxide synthase (an enzyme involved in the breakdown of Nitric Oxide) mRNA* expression via the extract.

Benefits

- Aid cardiovascular health
- Lower LDL cholesterol
- Antioxidant
- Vasodilator

Side Effects

- Upset stomach
- Diarrhea
- Gas

Dosage

- 300-640 mg of leaf extract 3x a day
- 1-4 grams dried leaves 3x a day
- 1 teaspoon of liquid extract mixed with water

WARNING

Please consult with a medical professional prior to administering Artichoke Extract. There is concern that Artichoke Extract may cause an **allergic** reaction if you are diagnosed with an allergy to the ***Compositae/Asteraceae*** family, such as ragweed, marigolds, daisies, etc. Those with **bile duct obstruction** or **gallstones** can exacerbate their condition as artichoke extract can increase bile flow.

Ashwagandha *(Withania somnifera)*

This plant can also be called Indian Ginseng, Poison Gooseberry or Winter Cherry and is part of the nightshade family. It has been used in traditional Ayurvedic medicine for the past 4000 years. It's an evergreen plant that has red fruits and small green flowers and is native to India, Yemen, and South-Eastern China. Currently, it has been sold and marketed as an *anxiolytic (anti-anxiety)* and an *adaptogen (stress adaptor)*; some short-term studies support stress or sleep effects, but use alongside chemotherapy is not established and needs oncology review.

Currently, the actual mechanism of action isn't fully known; however, the main bioactive components are *withanolide* compounds; the following are some of the active *withanolide* compounds.

- *In laboratory research, Withaferin A* has been seen to interact with the protein *Vimentin. Vimentin* has been seen to positively interact with wound healing, *angiogenesis (recycling/creation of blood)* while negatively affecting cancer growth and spread. *Withaferin A* can interfere and bond with *Vimentin*, this bonding can cause *Vimentin* to break apart. Low dosages of *Withaferin A* can have a selective effect on *Vimentin*, while higher dosages can have the same effect with other "wanted" proteins. *Withaferin A* may also be able to inhibit *proteasomal (breaks down degraded proteins)* activity. This inhibition can be beneficial for dysfunctional *proteasomes*, as dysfunctional *proteasomes* can contribute to a myriad of diseases or cancers.
- *Withanolide A* has been seen to be able to bind to *NF-kB essential modulator (NEMO). NEMO* is an activator of *NF-kB (involved in inflammatory, immune and cell survival responses)*; by *Withanolide A* binding to *NEMO* it has been seen to reduce the activation of *NF-kB.* This can become beneficial for those with inflammatory conditions such as arthritis or reducing cell survival in cases such as cancer.
- *Glycowithanolides* seem to interact positively with *GABAergic (synapses that interact with **GABA**)*, which may be the reason for its *anxiolytic (anti-anxiety)* effects. Its antidepressant benefits seem to come from *alpha2 adrenergic*; however, its full mechanism isn't fully known.

Proposed benefits (evidence varies)

- Analgesic
- Antispasmodic
- Antioxidant
- Anxiolytic
- Immune booster
- Anti-cancerous
- Anti-inflammatory
- Adaptogen
- Antidepressant

Side Effects

- Rise in body temperature
- Diarrhea
- Nausea
- Upset stomach
- Vomiting
- Abdominal cramps

Dosage

Currently, there is no set dosage although 300-500 mg has been seen to provide benefit

WARNING

Seek medical advice before use. Ashwagandha may affect people with **digestive disorders** and **ulcers** as it can irritate the intestinal tract. Ashwagandha can enhance immune activity in those with **autoimmune diseases**. Those with a **thyroid disorder,** it can increase **thyroid** function. Rare liver injury is reported. Avoid during pregnancy; review thyroid, autoimmune and medication risks. Long-term safety is uncertain.

Astaxanthin (*3,3'-dihydroxy-β, β'-carotene-4,4'-dione*)

Astaxanthin is classified as a keto-carotenoid responsible for giving flamingos, shrimp, krill and some algae their reddish pigment. It has a similar structure to that of Vitamin A and has been seen to offer similar eye benefits, but with lower toxicity when compared to Vitamin A. Currently you can see astaxanthin advertised mainly as an antioxidant and an anti-inflammatory. The majority of the commercially bought astaxanthin is derived from the bacteria *Haematococcus pluvialis (a type of algae)*, as it turns out this form is more bioavailable than animal occurring astaxanthin.

Since astaxanthin is fat-soluble, eating it with food will increase its bioavailability. With the added benefit mainly coming from its ketone groups at the end, it is able to embed itself into a cell's membrane allowing it to exert antioxidant activity in the *intracellular (inside)* and the *extracellular (outside)* side of the cell. The antioxidant effect observed has been seen to promote blood flow through reducing *nitric oxide* signaling interference due to oxidative radicals; needless to say this preserves the *nitric oxide* signal. Additionally, this effect can benefit the membrane in red blood cells, as astaxanthin has been seen to reduce the oxidation that would otherwise damage the cell.

- It can also aid cardiovascular health by increasing *adiponectin. Adiponectin* is a protein hormone that is involved in the break-down of fatty acids as well as glucose regulation. This has resulted in the increased uptake of fatty acids and *triglycerides* into muscle tissue rather than glucose which can result in lower cholesterol levels.
    - This mechanism may be the reason for fat loss and enhanced stamina levels; due to the utilization of fatty acids used in the demand for **ATP** synthesis.

The antioxidant benefits may apply to muscle tissue as it may reduce oxidative damage after training, mainly as it has been seen to promote normal levels of antioxidant enzymes like **glutathione** and *superoxide dismutase.* The anti-inflammatory properties seem to come from its ability to inhibit *Cyclooxygenase-1 (COX-1); COX-1* enzymes are involved in inflammatory, clotting, and vasodilating signals. The overall mechanism seems to apply to skin as well, as it can maintain skin elasticity and reduce oxidation induced by UV damage.

Benefits

- Improved vision
- Cardiovascular health aid
- Fat loss
- Antioxidant
- Promote strength
- Skin rejuvenator
- Neuroprotector
- Enhance stamina

Possible Side Effects

- Reddish skin tone

Dosage

- Current optimal dosage is not known; 6-15 mg have shown benefit, up to 40 mg have been reported to be used by athletes without toxicity

WARNING

Please consult with a medical professional prior to administering Astaxanthin. There is concern that Astaxanthin may affect those diagnosed with diabetes as it may affect blood sugar levels, those with **allergies** to **shellfish**, shellfish derived astaxanthin may cause an allergic reaction.

Astragalus

The name of Astragalus has two different variations *membranaceus* and *propinquus.* Each is synonymous with each other and does not refer to a different herb. Astragalus is made up of over 3,000 different species of herbs and shrubs, varying from vines to pale flowers all belonging to the *legume (bean)* family. *Astragalus propinquus* has been used in traditional Chinese Medicine and has been seen to have anti-inflammatory, cardiovascular, and longevity benefits. TA-65 is a proprietary product associated with cycloastragenol, a compound related to *astragaloside IV* and has been claimed to "age-reverse" the immune system by stopping biological aging and cytotoxic *T-cell (lymphocyte)* death. Human age reversal is unproven; NCCIH finds insufficient reliable evidence for any health condition.

The extract TA-65 has been seen to improve skin condition, mainly *keratinocytes (cells in the outer layer of the skin).* It can accomplish this by activating and increasing *telomerase* activity in *keratinocytes. Telomerase* is the mechanism that replaces the little bits of DNA that are lost in cell division which can result in damaged DNA or fused chromosomes. Telomere maintenance can influence replicative lifespan, but does not establish unlimited safe cell division or human rejuvenation. This mechanism naturally slows as we age; the benefit seems to apply to other cells too.

- The study done behind this may be uncertain as it is not sure whether or not those that sourced the study may have profited from this; additionally, it is believed that the entire Astragalus extract may promote *telomerase* to a higher affinity due to the other bioactive compounds.

This mechanism may extend to other areas of the body as Astragalus has shown protective effects on neurons and cardiac tissue, especially after damage from *ischemia (restriction of blood/oxygen to tissues).* Another believed mechanism may extend to antioxidant capabilities, although the exact mechanisms are currently unknown.

- It is also believed that Astragalus can relax and dilate arteries through the *Cyclic Guanosine Monophosphate (cGMP) Pathway. cGMP* is involved in smooth and vascular muscle relaxation; activation can lead to improved blood flow and vasodilation.

Proposed benefits (evidence varies)

- Anti-aging
- Cardiovascular health aid
- Adaptogen
- Antioxidant
- Vasodilator
- Immune booster
- Skin enhancer

Possible Side Effects

- Low blood pressure

Dosage

- Traditional decoction amounts are not equivalent to extract doses. No established dose reverses aging, and combining herbs needs individual review.

WARNING

Seek medical advice before use. Astragalus may interfere with **auto-immune diseases** as it can stimulate the immune system. Astragalus may affect Diabetics and those with blood pressure problems as it affects blood sugar and blood pressure. Astragalus is considered a **legume** and may affect those with **peanut allergies**. Research shows positive effects, though this herb is new to the market and much of its mechanism isn't fully known. Be wary of astragalus extracts; several species contain the alkaloid Swainsonine which has been known to cause neurological disorders.

Bacopa monnieri

Bacopa monnieri, also commonly known as water hyssop or Brahmi, can be seen primarily sold as a cognitive booster. Bacopa monnieri is an herb that originates in eastern India but can be found growing throughout various continents such as Europe and the United States. It's a green herb with white-ish small flowers, with the capabilities of being able to grow in water or marshy areas. This herb has been used in Ayurvedic medicine to treat a long list of medical problems such as ulcers and epilepsy.

Bacopa monnieri use in nootropics is mainly because it can promote the growth of *dendrites (extensions of nerve cells that connect/communicate with other neural cells),* in addition to its interaction with *dopamine (responsible for motivation/reward behavior)* and *serotonin (plays a part in learning and memory).* Expect a minimum of 4 weeks to notice any improved mental function as it needs to build up in the body. The main bioactive compounds are *bacosides*, the main ones of interest being *bacoside A* and *B* and its metabolites since they are able to cross the blood-brain barrier.

- *Bacoside A* has been seen to contain neuro antioxidant properties as it can promote levels of the antioxidant **glutathione** while also increasing enzyme activity like *Superoxide Dismutase (SOD)* and *glutathione reductase (GSR)* in the brain. Both of these enzymes play a role in resisting oxidative stress.

This mechanism is believed to reduce the risk and the effect of neurodegenerative diseases like dementia and Alzheimer's, as oxidative stress may be linked to the state of these diseases and symptom severity.

- Currently, although not fully understood, it has been seen that *bacoside* compounds are able to release *Nitric Oxide* from the *endothelium (the lining of blood vessels)* thereby increasing cerebral blood flow; this is beneficial as the majority of cerebral diseases such as dementia have been linked to a lack of blood flow in the brain.
- *Bacosides* are able to promote higher levels of *tryptophan hydroxylase (involved in serotonin creation)* and *serotonin transporter (transports and recycles serotonin)* during times of stress.

Benefits

- Cognitive enhancer
- Aid neurological problems
- Anxiolytic
- Adaptogen
- Antioxidant

Side Effects

- Increased bowel movements
- Fatigue
- Stomach cramps
- Nausea
- Dry mouth

Dosage

- Leaf powder form, it should be consumed with a fatty substrate like butter as it needs a lipid transporter to enter the body, generally 750-1,500 mg.
- 300 mg of a standardized extract of 55% bacoside content

WARNING

Please consult with a medical professional prior to administering Bacopa monnieri. There is concern that Bacopa monnieri can affect those diagnosed with **Bradycardia** as it can slow heart rate. Increased thyroid and lung secretions have also been noted which could worsen conditions such as **asthma** and possibly interfere with existing **thyroid conditions**. Bacopa monnieri can increase stomach secretions affecting those with **ulcers**.

<u>Banaba *(Lagerstroemia speciosa)*</u>

Banaba, from the *Lythraceae* family where pomegranates can be found, can go by the name of Giant Crape-myrtle or Queen's crape-myrtle. The Banaba tree can grow up to 66 ft and has been described as a small to medium-sized tree bearing purple-ish to white flowers with oval leaves. It is native to Asia and parts of the Philippines, and its leaves have been used as vegetables or brewed in tea. Traditionally, the leaves have been allowed to age and mature and used to control blood glucose levels.

The mixture of the bioactive compounds in Banaba leaves has been found to be *Corosolic acid* and *Valoneaic acid* amongst various other *phenolic* compounds. These compounds have been seen to inhibit the absorption and breakdown of carbohydrates in the digestive tract, although not entirely, mainly through inhibition of glucose transport.

- *Valoneaic acid* has been seen to inhibit the enzyme *Alpha-Amylase (breaks down glycogen and starches into glucose).*
- *Corosolic Acid* has been seen to inhibit *Alpha Glucosidase (involved in breaking down carbohydrates and starches into glucose).*

The inhibition of these enzymes has been shown to effectively inhibit the absorption of glucose and carbohydrates into the body, thereby resulting in lower blood glucose levels.

- *Corosolic acid* or *Gallotannins (currently, it's unknown which has this effect)* has been seen to be able to activate *Glucose Transporter Type 4 (GLUT4)* and transport glucose into various types of cells, to include muscle tissue and fat tissue.
- The combination of *ellagic acid* and its derivatives, like *3-O-methyl-ellagic acid 4'sulfate,* have also been seen to reduce *adipocyte (fat cell) proliferation (multiplication).* With the added benefit that they are *phenolic* compounds; therefore, they also have antioxidant properties; this also applies to *Corosolic Acid* as it is able to scavenge free radicals reducing oxidative damage.

Interestingly enough, Arjunolic *acid,* another active compound, has been seen to have anti-inflammatory properties as it can inhibit *cyclooxygenase-2 (creates inflammatory messengers like Arachidonic Acid)* activity. Other *phenolic* compounds have shown that they are able to inhibit the activation of *NF-kB (responsible for inflammatory actions).* The extract can help any athlete by promoting glucose transportation into muscle cells, thereby enhancing recovery.

<u>Benefits</u>

- Antioxidant
- Antidiabetic
- Anti-obesity
- Promote muscle growth
- Analgesic
- Anti-inflammatory

<u>Possible Side Effects</u>

- Low blood pressure

<u>Dosage</u>

- 3,000 mg leaf extract with a standardized *corosolic acid* content (adhere to label instructions)

<u>***WARNING***</u>

Please consult with a medical professional prior to administering Banaba. There is concern that Banaba can affect those diagnosed with **Diabetes** as Banaba has been shown to lower blood sugar levels and increase insulin sensitivity; medication might need to be modified. If you have been diagnosed with **low blood pressure,** Banaba has been shown to lower blood pressure to unsafe levels.

Barley Grass *(Hordeum vulgare)*

Barley is a part of the grass family, *Poaceae*, like wheat and is believed to be one of the first grains that were cultivated in Europe and Asia. It has a long use in the creation of bread, foods, beer, and other distilled drinks. Barley grass is the seedling of the plant that is generally harvested 200 days after it germinates. It can be seen used in powder and has been a go-to for the various fitness enthusiasts and health advocates ranging all the way from day-to-day gym-goers to nutrition experts. Barley grass is nutrient-rich containing all 9 essential amino acids along with Vitamins A, C, B1, B2 and folate; it is not a dependable source of active B12. It also contains **GABA**, calcium, **iron**, **potassium** and most importantly, **chlorophyll**, which is believed to be where some of the most profound health benefits of Barley Grass come from.

- GABA can modulate and reduce neurotransmissions through balancing the neurotransmitter **glutamate**. This can prevent neural overexcitability and neural oxidation which can lead to proper cognitive function, and since GABA plays a role in sedation and relaxation, it can aid in regulating sleep and stress, as well as exert anxiolytic effects.
- Barley Grass has also been seen to have various antioxidant enzymes such as SOD, an antioxidant that is able to convert the *Superoxide Radical* (O_2^-) into oxygen and hydrogen peroxide; Superoxide *Radicals* are toxic and can cause oxidative damage in the body.
- *Chlorophyll* is the pigment plants use to capture light energy. It shares a ring framework with heme, but contains magnesium rather than heme's iron and is not an oxygen-carrying protein. Laboratory studies of chlorophyll derivatives explore carcinogen binding, oxidative signaling and cancer-cell growth. These findings do not mean that consumed chlorophyll replaces *hemoglobin*, oxygenates tumors or treats cancer. Cancer cells can also use oxygen; their altered metabolism is more complex than an exclusively anaerobic state. Clinical an*ticancer benefit from this supplement is unestablished.*

It has been seen to decrease *lipid peroxidation* of *Low-Density Lipoprotein (LDL)* through its antioxidant effect; this can reduce the *atherosclerotic* build-up (plaque build-up on the arteries). Oxidized *LDL* can attach itself to *macrophages (white blood cells)* causing them to become *foam cells. Foam cells accumula*te in artery walls. By decreasing oxidation, the risk of *foam cells* decreases as well as the risk of cardiovascular disease.

Proposed benefits (evidence varies)

- Anti-inflammatory
- Antioxidant
- Antidiabetic
- Cardiovascular health Aid
- Anti-Arthritis
- Anti-cancerous

Possible Side Effects

- Laxative effect

Dosages

- Adhere to label instructions; if juicing, 100 g of barley grass is generally added to juices

WARNING

Seek medical advice before use. Barley Grass can affect people with **Celiac disease** or any gluten sensitivity/allergy as Barley Grass/powders can contain gluten making symptoms worse. Barley Grass can trigger allergic reactions in those with **allergies** to **grains**. Barley Grass can lower blood glucose levels affecting **Diabetics.**

Beta-Alanine (CarnoSyn®)

Beta-alanine is a naturally occurring amino acid that is a rate-limiting precursor to **carnosine**, which basically means that the Beta-Alanine levels in the body are what limits the further synthesis of carnosine. The increase of carnosine in muscles is where the performance benefits of Beta-Alanine—fatigue reduction and increased total muscle tissue activation—are derived from. While the body is able to create its own Beta-Alanine, it is not able to increase carnosine levels in muscle tissue in the same fashion as supplementation can, mainly due to the fact that we do not consume as many precursors needed to achieve the same concentrations. With that being said, it is important to note that carnosine supplementation alone will not increase concentration levels in the muscle, nor is it as effective as beta-alanine supplementation. Benefits are mainly seen in areas of exertion lasting 60-240 seconds.

The basic mechanism of Beta-Alanine comes after it is converted into carnosine by attaching a **histidine** amino group via the *ATP-Grasp Domain-containing protein 1.* Carnosine can be abundantly found in skeletal muscle tissue, Type I and Type II muscle fibers, and it's released when the pH levels start to drop in an attempt to buffer acidity (aka fatigue). It is currently believed that it accomplishes this by maintaining a balance of *Hydrogen (H^+) Ions.*

- During exercise, lactate becomes a byproduct of the *Anaerobic Glycolysis System.* This is the process where glycogen is broken down to glucose and used for energy instead of oxygen, resulting in Lactic acid creation. L*actate dehydrogenase converts pyruvate to lactate while regenerating NAD+ and consuming H+. Acidity during intense exercise is* linked to high ATP turnover, not lactate simply releasing *acid.*
- This process will continue until the amount of H+ Ions has accumulated to the point of creating an environment where *Glycolysis* is inhibited, and muscle contractility is diminished; carnosine can combat this by resisting a pH level change by neutralizing small acid changes in the cell.

Studies have shown an increase in strength/stamina vs. fatigue threshold. Since it serves as a buffer for fatigue, strength may not generally be increased by just supplementing alone; rather it allows the user to stimulate muscle tissue past the point of fatigue allowing the muscle fibers to achieve more powerful contractions.

Benefits

- Increase anaerobic capacity
- May improve strength
- Fatigue reduction
- Promote lean muscle
- Increase endurance
- Antioxidant
- May increase strength (When coupled with creatine)

Possible Side Effects

- Paresthesia (The prickly feeling you get on your hands and face after ingestion)
- Flushing

Dosage

- Studied regimens commonly use 4–6 g/day in divided doses for several weeks. Tingling is dose-related; routine taurine co-supplementation is not established as **ne**cessary.

WARNING

Seek medical advice before use. Beta-Alanine can affect those with pre-existing **taurine deficiencies** as they use the same transporter, *SLC6A6 uptake transporter*, increasing the risk of taurine imbalance.

Betaine Anhydrous *(Trimethylglycine [TMG])*

Betaine is an amino acid that can be found in some foods such as beets (where it was originally discovered) or wheat germ. It is currently marketed as a potent cell protector from stress; this benefit has been observed in plants and humans, mainly because it is an *osmolyte. Osmolytes* can positively affect *osmosis* around a cell, maintaining proper cell fluid and volume resulting in proper hydration status. Promoting cell hydration increases resilience to stress. It has also been seen as a *methyl donor*; *methylation* plays a huge role in various processes in the body from the creation of neurotransmitters to RNA processing. Faulty *methylation* can cause various abnormal conditions in the body like Chronic Fatigue Syndrome.

Betaine has been seen to be able to increase cell *tonicity (osmotic pressure measurement)* by increasing fluid capacity potential and volume within the cell, similar to the effect seen in **creatine**. It is able to accomplish this as it can be transported in or out of the cell through the *betaine/GABA transporter 1,* due to its *osmolyte* properties.

- This is where, in performance, benefits have been observed as it can preserve cells and increase survivability against the damaging stress of exercise. (Since exercise creates *Hydrogen Ions* which can damage cells.)

Betaine also has cardioprotective properties as it has a high affinity for cardiac tissue. This is mainly due to its ability to maintain healthy **homocysteine** levels. Betaine is able to donate a methyl group to homocysteine, allowing it to convert into **L-Methionine**.

- High levels of homocysteine have been implicated in the activation of the enzyme *elastase (breaks down elastic fibers known as elastin).* This can compromise and degenerate the integrity of arterial tissues/membranes. Homocysteine can additionally induce **collagen** production in smooth muscle cells leading to the creation of *fibrosis* and plaque build-up within arterial walls. This causes homocysteine to further activate *cyclins*; *cyclin* is a protein involved in cell *proliferation (cell multiplication)* increasing the risk of heart disease.
- Betaine's conversion of homocysteine to methionine can promote creatine synthesis in the body as the provision of methionine is able to produce **S-adenosyl methionine**, which is needed for creatine synthesis.

Benefits

- Increased muscle growth
- Increased endurance
- Increased strength
- Cardiovascular health aid
- Adaptogen
- Antioxidant

Side Effects

- Lasting unpleasant taste
- Body odor
- Nausea
- Upset stomach
- Diarrhea

Dosage

- 1,250-1,500 mg twice a day

WARNING

Please consult with a medical professional prior to administering Betaine Anhydrous. There is concern that Betaine Anhydrous can affect those diagnosed with **high cholesterol** as it has been seen to increase cholesterol in individuals.

Bilberry *(Vaccinium myrtillus)*

Bilberries are a separate berry, and outside of the small changes in the name, you would be confused and hard-pressed when you realized that they are not blueberries. These berries look almost identical to blueberries, and for a good reason, they are within the same *genus*, *Vaccinium*. They are native to Europe while blueberries are native to North America. *It's important to note that the antioxidant anthocyanin is difficult to attain through food consumption alone. Less than 5% of anthocyanins remain after the fruit is digested; the remains are metabolites of anthocyanin, and they are excreted rapidly by the body.* Bioavailability is increased when taken as *liposomal micelles (these are a thousand times smaller than a human red blood cell)*; this is currently undergoing cancer research. This does not establish that concentrated extracts improve night vision or treat eye disease; human evidence is insufficient.

Bilberries' main bioactive compounds would be *anthocyanins*. *Anthocyanins* have been shown to have antioxidant properties against oxidation and protection against *abiotic* (non-living materials water, light, radiation, etc.) insults, making it beneficial for UV damage/protection. *Anthocyanins* have also been shown to promote blood vessel health and circulation, which would enhance vasodilation and may also lower blood pressure.

- *Anthocyanins* have been seen to inhibit *α-glucosidase (involved in digesting carbohydrates and starches into glucose)*. *Cyanidin-3-galactoside* in specific can work synergistically with *acarbose (antidiabetic medication)*. These mechanisms can, independently or with medication, lower blood glucose levels.
- *Anthocyanin* has been shown to reduce levels of pro-inflammatory *cytokines (cell messengers)* such as *Interleukin-6 (IL-6)* as well as maintaining proper *Interleukin-15 (IL-15)* responses. *IL-15* has been implicated in obesity-related inflammation and diminished glucose function as it regulates activation and *proliferation (cell multiplication)* of *natural killer (NK)* cells, specifically *CD56^{bright} NK* cells which have been seen to mainly produce proinflammatory *cytokines*. *Variant NK* cells have also been seen to maintain glucose homeostasis by producing *IL-13* which modulates *gluconeogenesis (generation of glucose)*.

The anti-hypertension/vasodilator effects are currently not known, although it is believed that this effect is mediated through its anti-inflammatory effects as an increase of the previously mentioned inflammatory markers has been linked to increased/high blood pressure.

Proposed benefits (evidence varies)

- Anti-inflammatory
- Antioxidant
- Nootropic
- Antidiabetic
- Vasodilator
- Anti-cancerous

Side Effects

Currently, there have been no clinically/anecdotally reported side effects; data is still lacking.

Dosage

- Currently, there's no data stating a recommended dosage; adhere to label instructions

WARNING

Seek medical advice before use. Bilberry extract can affect those **Diabetics** as it interacts with glucose levels and transport mechanisms.

Bio-Active Peptides

Bio-Active Peptides and regular peptides, in general, are just short chains of amino acids linked by peptide bonds. A peptide can be a *dipeptide (2), tripeptide (3), tetrapeptide (4),* etc. Generally, the source of the name is based on chain size; polypeptides have been referred to as proteins, while smaller peptides like the ones mentioned above are seen as amino acids. However, there is no set rule about this. It is important to note that the process used to acquire peptides changes their use and benefit. There are antidiabetic peptides, antihypertensive peptides, antimicrobial peptides, growth hormone peptides, etc. Bio-Active Peptides, however, are like the peptides explained above, except that they differ in design and purpose. A peptide's design and purpose, for example, could be a growth hormone which is made up of a 191 amino acid sequence. Now let's say the delivery of the growth hormone peptides wasn't administered correctly; the stomach acid would break it apart into smaller fragments, losing its design and becoming basic amino acids used for dietary purposes. For the sake of space, Bio-Active Peptides will be discussed on this page as it is a commonly known supplement.

- Proteins (like the powder you drink) are seen to consist of one or more polypeptides that are about 50 amino acids long.

Bio-Active Peptides' design is derived from a particular combination of **Colostrum** components;

- *Proline-Rich Peptides (Colostrinin)-* A neuroprotector and immune booster as it is able to inhibit the overproduction of *Reactive Oxygen Species (responsible for oxidative damage).* It is additionally able to regulate *T helper Cell 1* and *2.* These cells help modulate immune responses and maximize the bactericidal activity of *phagocytes (cells that eat bacteria and viruses.)* It has also been seen to positively regulate *neurite (either an axon or dendrite)* growth.
- *Fibroblast GF-* Basically a growth factor, stimulates the production, regeneration, and repair of tissue. *FGF 2* and *6* specifically are able to regenerate muscle. *FGF 2* can recruit *myosatellite cells (can differentiate into a muscle cell).* High *FGF6* levels have been linked to muscle regeneration.
- *Lactoferrin-* Transports iron ions while additionally acting as a protector of oxidative damage at a molecular level. Moderate oxidation can cause *apoptosis (programmed cell death)* while severe oxidation can cause *necrosis* of tissues. It is able to sequester iron preventing metal-based *lipid peroxidation.* It is also able to reduce *apoptosis* by reducing *caspase-3 (involved in executing apoptosis)* activation.

Benefits

- Promote muscle growth
- Promote strength
- Enhanced recovery
- Enhance immune function
- Aid injury repair
- Antioxidant

Side Effects

- Upset stomach
- Bloating
- Gas

Dosage

- No set dosage has been set, adhere to label instructions

WARNING

Please consult with a medical professional prior to administering Bio-Active Peptides. There is concern that Bio-Active Peptides can affect those diagnosed with **lactose allergies/sensitivities** as it may cause an allergic reaction. It's important to note that there are secretagogue peptides, which can affect those with **hormone conditions** or those undergoing **hormonal therapy**.

Biotin (Vitamin B7)

Most people recognize biotin as a supplement for hair, nails, and skin; this does carry some merit as Biotin deficiency symptoms are generally characterized by thinning hair or skin rashes. Biotin is an enzyme *cofactor (needed to complete enzyme processes)* needed for various process in the body like in the conversion of energy since Biotin aids in the conversion of the food and helps convert it into a usable form. Foods such as eggs, liver, cheese, peanuts, and **wheat germ** tend to be high in biotin.

Biotin is a water-soluble **B Vitamin** and has been seen to be involved in biotin-dependent metabolic processes like fatty acid synthesis, BCAA catabolism, and the generation of glucose. In simpler terms, it breaks down fats, carbohydrates, and proteins into energy. One of the enzymes being *acetyl-CoA carboxylase,* this enzyme plays a role in the synthesis of *fatty acids* and prevents their oxidation. Another is *3-methylcrotonyl-CoA carboxylase*; this enzyme plays a role in creating *acetyl CoA* and breaking down **Leucine.**

- *Acetyl-CoA* is also used in the *Citric/Krebs Cycle* to release stored energy; this is accomplished through oxidizing *acetyl-CoA,* which is oxidized to CO2, reducing electron carriers, thus creating energy. *CoA's thiol group forms thioesters such as acetyl-CoA;* thioesters have a high energy bond that happens to be exergonic (dissipates to the surrounding area); the energy released is then used for **ATP** and *GTP (GTP is used for RNA synthesis).*

Biotin can also interact with *glucokinase*, an enzyme known to play a role in the cellular use of glucose as it can regulate carbohydrate breakdown and use in response to glucose levels in the blood.

- There really isn't a need to supplement Biotin for general health and metabolic functions as it can easily be acquired through diet and some multi-vitamins. Benefits for hair, skin or nails without deficiency are not established.

For those that may have a Biotin deficiency, symptoms can include depression, fatigue, tingling in the arms and legs, dry skin, and seborrheic dermatitis in infants.

- Seborrheic Dermatitis is a skin condition that causes flaky and scaly patches along with red skin mainly found on the scalp but can be seen elsewhere on the body.

Benefits

- Increased energy
- Skin health

Possible Side Effects

- No common side effects have been reported through regular consumption of Biotin

Dosage

- Recommended dosages are as follows; depending on the brand, follow label instructions.
  - 10 micrograms daily 1-8 years of age
  - 20- 25 micrograms daily 8-18 years of age
  - 30 micrograms daily 18 and older

WARNING

Seek medical advice before use. Biotin can interfere with laboratory tests, including some troponin and thyroid assays, producing misleading results. Tell the clinician and laboratory about use before testing; stopping intervals depend on the dose and test.

Black Cohosh

Black Cohosh, also known as Black Snakeroot and Fairy Candle, is native to North America. It has been used in traditional Native American medicine since the extract has been seen to contain *analgesic (pain reducer)*, sedative, and anti-inflammatory properties. The roots have been known to contain the most bioactive compounds found in the plant. Currently, it has been mainly studied as an alternative treatment as a menopause symptom aid, particularly aimed at hot flashes and night sweats.

The main bioactive compounds found in Black Cohosh are *triterpene glycosides,* specifically *acteol, cimigenol, 27-deoxyactein,* and *cimicifugoside.* These compounds have been seen to interact in a *serotonergic (pertaining to serotonin functions)* fashion instead of estrogenic functions like previously believed. If anything, it seems to be an anti-estrogenic compound as it can inhibit breast cancer cell growth which is dependent on estrogen to grow; it also seems effective at suppressing *Luteinizing* hormone secretion; however, its mechanism of action is still not yet known.

- *N-methylserotonin,* a *tryptamine alkaloid,* has been recently found in Black Cohosh. It has been seen to interact with the *thermoregulatory (body heat regulator)* $5\text{-}HT_7$ *receptor (a serotonin receptor).* Activation of this receptor can also have an antidepressant effect and may be related to its anti-menopausal effects. It has also been seen to activate *mu-opioid receptors* which may be related to its analgesic effect.

Current studies and research have not found a direct or concrete mechanism of action as various parts and components of the extract seem to contradict each other. As of now, it is believed that its effects can still be related back to Black Cohosh being a *Selective Estrogen Receptor Modulator (SERM),* meaning it can be an estrogen *agonist (activator)* or *antagonist (blocks activation)* dependent on the tissue/area, positively modulating estrogen in the body.

- It's important to note that not all the pharmacological effects are fully understood and more data and research are needed to have a conclusive explanation for the molecular interactions.

Proposed benefits (evidence varies)

- Aid menopause symptom
- Menstrual symptoms aid
- Bone density aid

Side Effects

- Dizziness
- Headache
- Nausea
- Vaginal spotting
- Low blood pressure

Dosage

- 1-2 mg of triterpenoid glycosides should be included per dose regardless of the type of extract; it should state on the label if the product is legitimate. Adhere to label instructions.

WARNING

Seek medical advice before use. Long-term use may cause thickening of the womb lining; this worry arises amongst doctors as they believe it could increase the risk of womb cancer. Some individuals have reported liver damage while using Black Cohosh, including serious cases; causation is uncertain, and symptoms of liver injury require stopping use and medical care. **DO NOT USE BLACK COHOSH WHILE PREGNANT AS IT MAY INCREASE THE RISK OF MISCARRIAGE.** Its hormonal actions and long-term safety remain uncertain; do not consume black cohosh if estrogens may affect any pre-existing conditions.

Borage Oil *(Borago officinalis)*

Borage Oil is derived from the seeds of the *Borago officinalis* plant which can also be known as the Starflower. Borage is an annual herb that is native to the Mediterranean region; the plant can also be seen growing in Europe. The leaves and flowers are edible, the leaves tasting like cucumbers and the flowers having a sweet honey-like taste. The plant does contain *pyrrolizidine alkaloids.* **It is important to note that Pyrrolizidine alkaloids can be carcinogenic and toxic to the liver. Pyrrolizidine may be present in extracts but can be removed through processing; oral products must be certified free of toxic unsaturated pyrrolizidine alkaloids.** Currently, it can be seen marketed as a therapeutic aid to various degenerative diseases like *rheumatoid arthritis,* as well as a skin enhancer and an aid for various other inflammatory conditions.

The major mechanism of the plant comes from its polyunsaturated fatty **omega-6** content. The seeds contain 26-38% borage oil, which contains 17-28% **gamma-linolenic acid (GLA),** an omega-6 fatty acid. Healthy adults are able to create a lower amount of GLA within their bodies from dietary *linoleic acid; this does not establish a general need for borage* supplementation.

- GLA is converted by *elongation* into *dihomo-y-linolenic acid (DGLA),* a precursor to *prostaglandins (A compound that has hormone-like effects found in almost every tissue in humans)* and prevents *leukotriene* synthesis. It appears to accomplish this as *DGLA* can be used in cellular *phospholipids* that can later be activated and converted by *Phospholipase A2* in the cell; this releases metabolites with anti-inflammatory properties such as *prostaglandins.*
    - *Prostaglandins* can also act as powerful local vasodilators
    - *Leukotrienes* are inflammatory mediators produced by *leukocytes (white blood cells)* from the oxidation of **Arachidonic Acid; DGLA** has been seen to also have competitive synthesis against Arachidonic Acid, which could ultimately reduce inflammatory metabolites.

This mechanism may additionally have anti-thrombotic effects (anti-blood clotting) benefits. It's important to note that many people like to consume CLA for fat loss effects. Do not confuse GLA with **CLA**. Neither a proposed pathway nor replacing oils establishes clinically meaningful weight loss from GLA or CLA.

Proposed benefits (evidence varies)

- Cardiovascular health aid
- Vasodilator
- Anti-inflammatory
- Joint flexibility
- Healthy triglycerides
- Healthy skin

Side Effects

- Bloating
- Diarrhea

Dosage (Dosages can vary dependent on extract adhere to label directions)

- 1-2 grams of a standardized extract for eczema daily
- 2-3 grams of a standardized extract for rheumatoid arthritis daily
- 200-400 mg of a standardized extract for breast pain daily

WARNING

Seek medical advice before use. Borage Oil can affect people with **bleeding disorders** as Borage Oil can reduce clotting. Borage Oil has been seen to affect blood sugar levels; **Diabetics** should adjust according to sugar levels.

Boron

Boron is a chemical element found on the Periodic Table sitting at atomic number 5. It is in the Metalloid group, where Silicon and Arsenic can be found. This group has properties somewhat in between metals and non-metals. Boron can be found in various states and is mainly mined as an *evaporite (sedimentary rocks formed through crystallization by evaporation)* since it is water-soluble. Boron is considered a micronutrient and can generally be consumed in small trace amounts through fruits, vegetables, and legumes.

Boron seems to interact with the production of testosterone and estrogen. At 3 mg it was found to raise testosterone levels which might aid at increasing muscle mass while at 2.5 mg, it seemed to raise estrogen levels in women. Further research needs to be done to establish conclusive dosages. Its mechanism may come from its ability to increase the half-life of the mentioned hormones by inhibiting *microsomal enzymes* that catabolize those hormones; this is the theoretical belief; however, its full scope and mechanism isn't fully understood.

Boron has been seen to interact with *osteogenesis (bone formation)* as it can positively regulate genes involved in the mineralization of *osteoblasts (cells that create bone matrix)*. It has also been seen to positively modulate hormones involved in bone growth such as testosterone and *17β-estradiol.* Boron has also been seen to enhance Vitamin D utilization in the body, which is important as Vitamin D stimulates calcium absorption in the intestines, which can promote the highest affinity for bone mineralization. It has also been proposed, but data is inconclusive, that Vitamin D can inhibit *osteoclasts (cells that break down bone tissue)* thereby promoting mineralization of bone tissue. This mechanism can result in more dense and solid bones.

- A similar mechanism has been seen for cells found in connective tissue as it enhances *fibroblast* enzymes that synthesize *extracellular-matrix (molecules outside of the cell that provide support and structure to cells)* and **collagen**.

Proposed benefits (evidence varies)

- Anti-osteoporosis
- Improve bone density
- Menopause aid
- Improve wound healing
- Promote healthy hormone levels

Side Effects

- Nausea
- Vomiting
- Abdominal Pain
- Rash
- Convulsions

Dosage

- Research showed 1-3 mg per day was found safe, although 2.5 mg can cause hot flashes and night sweats in women; therefore, women going through menopause might want to avoid Boron use.
- Testosterone and performance benefits are not established. No human RDA exists; the adult upper limit is 20 mg/day, a safety ceiling rather than a target.

WARNING

Seek medical advice before use. Boron can affect people with **Hormone Sensitive Conditions** such as **Prostate**, **Ovarian,** and **Breast Cancer** as it can affect estrogen and testosterone levels. If you are diagnosed with **kidney disease** boron can increase kidney activity.

Boswellia Serrata *(Salai guggul)*

Boswellia has long been used in religious ceremonies as an incense (Frankincense), mainly for its fragrant resin; only four Boswellia species can actually bear Frankincense. Boswellia generally comprises medium-sized trees and shrubs that are native to Eastern Asia and Northern Africa. The tree's sap has been historically used in traditional Ayurvedic Medicine in India. In Europe, it has currently been classified as a *nutraceutical (pharmaceutical of plant origin).* It has been seen to have anti-inflammatory and *analgesic (pain reducer)* effects and some trials suggest symptom relief in osteoarthritis, but study quality and preparations vary.

The active components of the extract are what are known as *Boswellic acids,* mainly *11-keto boswellic acid (KBA)* and *3-O-acetyl-11-keto-β boswellic acid (AKBA).* These acids make up to 30% of the resin, and it is important to note that these acids are non-volatile, and therefore, the burning of the frankincense or the extracts will not provide any effect. Additionally, the acids are too heavy to be carried by steam; therefore, hydro-distilled frankincense oil or the extracts can not provide the same effect. The acids are also hydrophobic (fat-soluble) not hydrophilic (water-soluble).

- Boswellia's main inflammatory mechanism seems to come from AKBA's ability to inhibit *NF-kB (involved in cell signaling)* activation from the various inflammatory mechanisms, not just one. It does this by inhibiting the *phosphorylation (the addition of a phosphate group)* of the *Inhibitor of kappa B (IkB).* If this cannot happen, then *NF-kB* cannot enter into the nucleus where it can alter gene expression into inflammatory signals/responses.
    - *IkB* is always attached to *NF-kB,* which keeps it from entering into the nucleus. Once it detaches itself through *phosphorylation (this needs to happen for activation, like letting it off the leash in a way)* it is a permanent action and cannot be re-attached; this is why there is an inflammatory down cascade which can lead to or aggravate other inflammatory problems in the body as it takes time to re-synthesize *IkB.*

This mechanism may be the reason why *boswellic acids* have also been seen to reduce and inhibit the synthesis of *5-lipoxygenase, a* pro-inflammatory enzyme, which can ultimately lead to a reduction in inflammation-related pain.

Proposed benefits (evidence varies)

- Pain reducer
- Joint health aid
- Anti-arthritic
- Aid inflammatory conditions

Side Effects

- Gastrointestinal discomfort
- Diarrhea
- Nausea

Dosage

- Extracts can vary in potency, adhere to label instructions
- Resins are 1.2 g 3x a day while brand name extracts are around 100-250 mg due to the high concentration of boswellic acids.

WARNING

Seek medical advice before use. Boswellia Serrata can cause an **allergic reaction** to those with sensitive skin as it has been seen to cause a rash. It may also increase menstrual flow which can cause a miscarriage if **pregnant**.

Bromelain *(Ananas comosus)*

Bromelain is derived from pineapples and is currently marketed as a digestive aid. Prescription topical bromelain-based products remove dead burn tissue; this evidence does not apply to oral supplements. In culinary professions, it can be found used as a meat tenderizer mainly because it is a *proteolytic (breaks down proteins into peptides and amino acids)* enzyme.

The active components of bromelain are *Ananain* and *Comosain*; these enzymes are responsible for *proteolysis*. If supplemented with meals, it can aid with digestion as well as boosting the immune system.

- Bromelain's anti-inflammatory properties may come from its ability to remove (maybe digest) the receptors on immune cells that respond to inflammatory responses. This can ultimately lead to an anti-inflammatory state within the body.
- Bromelain has also shown possible cardioprotection as it has *fibrolytic (able to break down complex polysaccharides)* effects which can promote an anti-platelet, anti-adhesion environment and inhibit *thrombus (blood clot)* formation within the arteries.
- It has also been seen to prevent *foam cells; foam cells* are caused by *oxidized LDL* attaching to *macrophages (white blood cells). Foam cells accumulat*e in artery walls, which can lead to *atherosclerotic* build-up. Bromelain can eliminate the *CD36 (binds to oxidized LDL)* receptors on *macrophages*, thereby preventing *foam cells.*
- It has also been observed to reduce (probably through digestion again) the *CD44* receptor on cells' surfaces, and while important for various functions, *CD44* has been linked to tumor *metastasis (the spread of tumor cells to other areas of the body). CD44* plays a role in *angiogenesis (new blood vessel growth); tumor angiogenesis* is needed to be able to break through the *extracellular matrix* and go elsewhere in the body.
- Bromelain seems to promote positive immune modulation, activating an appropriate immune response rather than a chronic one, while additionally dampening any additional inflammatory stimulation. It has been seen to downregulate *Cyclooxygenase-2 (chronic inflammation)* while activating *Interleukin (IL)-1β, IL-6,* and *Tumor Necrosis Factor-α* pro-inflammatory mediators.

<u>Please note bromelain may self-digest, so it's advised to not leave it mixed in a drink for too long.</u>

Proposed benefits (evidence varies)

- Digestive aid
- Enhance the immune system
- Anti-inflammatory
- Reduce heart burn
- Decongestant

Side Effects

- Upset stomach
- Gas
- Diarrhea

Dosage

Data is fairly new adhere to label instruction

- Oral trials use varied preparations and doses; efficacy for many promoted uses and long-term safety remain uncertain.

WARNING

Seek medical advice before use. Bromelain can cause an **allergic reaction** in those with **allergies** to pineapples, latex, wheat, celery, papain, carrot, fennel, cypress pollen, or grass pollen.

Burdock *(Arctium lappa)*

Burdock is from the *Asteraceae* family where sunflowers and daisies can be found. It has been cultivated for years in Europe as a vegetable since the root is edible. In various other countries, like the United States, it is considered a weed more than anything since it can be found growing intrusively in nitrogen-rich soil. It can be seen marketed as a detox agent with touted claims like a blood detoxifier or a _______ detox (fill in the blank). Its main interaction seen within the body, however, is as an anti-inflammatory, antioxidant, hepatoprotection and as a possible vasodilator.

Some of the active components of this plant are *Arctigenin, Diarctigenin, Lappaol,* and *Inulin.*

- *Arctigenin* has been seen to be *cytotoxic (toxic to cells),* causing necrosis to cancer cells. It has also been seen to create an *anti-proliferative (cell multiplication)* environment for cancer cells, while in other circumstances, it has had *apoptotic (cell death)* effects on cancer cells. These effects are dependent on cancer cell types, and the effectivity will vary dependent on the cancer.
    - *Diarctigenin* has been shown to attenuate inflammatory conditions by inhibiting *Nitric Oxide* synthesis as well as reducing the production of inflammatory mediators *Interleukin (IL)-1β, IL-6* and *Tumor Necrosis Factor-α,* all of which are responsible for inflammatory responses. It can also reduce the DNA binding ability of *NF-kB*; this can directly affect inflammation since *NF-kB* is needed in the nucleus of a cell to transcribe inflammatory signals.
- *Lappaol* may also reduce the survivability rate of some cancer cells as it can amplify nutrient-deficient environments where tumor cells have difficulty receiving nutrients/glucose, thereby leading to cell death.
- Inulin has been seen to increase antioxidant enzymes in the body, such as *Super Oxide Dismutase, Glutathione peroxidase,* and *catalase. (These are endogenous antioxidant enzymes)*

Antidiabetic effects have been observed, although mechanisms need to be further assessed; as of now, it has been seen to increase insulin production via *beta cells (store and release insulin)* while reducing intestinal glucose absorption. The effect has been proposed to be from the antioxidant function found in the extract, thereby protecting and restoring pancreatic function where *beta* cells are found.

Benefits

- Antioxidant
- Anti-inflammatory
- Anti-cancerous

Side Effects

- Gastrointestinal discomfort
- Low blood pressure
- Contact dermatitis

Dosage

If using an extract, adhere to label instructions. The traditional dose was 2 g of root extract

WARNING

Please consult with a medical professional prior to administering Burdock. There is concern that Burdock may affect those diagnosed with **Diabetes** or have insulin sensitivities as Burdock can affect blood sugar levels, and medication may need to be modified. Please monitor blood sugar levels closely. If you are on **diuretics,** burdock can have diuretic effects as well. High consumption of tannins (antioxidants/polyphenols) have been shown to be toxic leading to upset stomach or *nephrotoxicity (kidney toxicity)*; burdock extract can be a good source of tannins and can cause this effect at high doses.

Conjugated Linoleic Acid (CLA)

CLAs are naturally occurring polyunsaturated omega-6 fatty acids that can be manufactured from linoleic-acid-rich oils such **as safflower oil**, although they can also be found in grass-fed beef. Currently, you can see CLA marketed as a fat loss aid that can increase lean muscle mass; you can find it on the label of some Amino Acids or weight loss products. The basis of these benefits is that theoretically, CLA is involved in the molecular signaling of fat burning, steroid, inflammation and glucose receptors.

The main mechanism seen in CLA's benefits is through its activation of the *Peroxisome Proliferator Activated Receptor (PPAR).* CLA is just *Linoleic Acid* with *conjugated bonds (the bonds increase stability in the molecule).* This means that there are various CLA molecules; however, only the *isomers (same compound with different molecular arrangement) trans 10, cis 12 (t10c12)* and *cis9, trans 11(c9t11)* have been shown to have more pronounced effects with each having its own unique effects. The isomers are able to bind and activate PPAR; PPARα is generally seen to be activated under conditions of nutrient deprivation and is used in fatty acid oxidation. This activation is necessary for *ketogenesis (the creation of ketones through the breakdown of fatty acids within the body.)*

Isomer t10c12 in CLA seems to be the more bioactive isomer, in regard to fat-loss, as it can inhibit PPARγ. Since PPARγ is involved in *lipogenesis (creation of fat), it* can actually reduce the size of the fat cell rather than reduce the number of fat cells. However, *PPARγ* inhibition also seems to temporarily increase insulin resistance.

- *Isomer t10c12* seems to have a higher affinity towards fat tissue, increasing lean muscle mass while reducing fat by transitioning to burn fatty acids instead of glucose for fuel.

Isomer c9t11 is antidiabetic, showing enhanced insulin sensitivity, while *isomer t10c12* can cause temporary inflammation of fat cells. This mechanism reduces additional glucose and fatty acid uptake into fat cells, while additionally increasing glucose and fatty acid uptake in muscle cells.

- Isomer *c9t11* tends to have a higher affinity towards muscle tissue; the mechanisms contradict each other, but it is where the fat-loss effects come from.

Proposed benefits (evidence varies)

- Increase in muscle growth
- Fat-loss

Side Effects

- Diarrhea
- Fatigue
- Upset stomach
- Nausea
- Headache

Dosage

- Weight-loss trials use varying doses; effects are small and inconsistent, with possible adverse metabolic effects.
- No established effective dose increases muscle growth in healthy adults.

WARNING

Seek medical advice before use. CLA may affect **Diabetics** as CLA has been known to affect insulin sensitivity/resistance. Those diagnosed with an **autoimmune disorder** are advised to talk to a medical professional as well as CLA may increase white blood cell count; it might interact with immune-suppressing medications. If you have **metabolic syndrome, taking** CLA might put you at risk for developing diabetes.

Caffeine *(1,3,7-trimethylxanthine)*

Caffeine has been globally marketed as an anti-sleep/stimulatory compound that can be found in energy drinks, pre-workouts and some teas. While the list of caffeinated products is vast, it is a commonly used substance. Caffeine is a *xanthine* like **Theobromine,** but unlike theobromine, it is a stimulant with the classification of nootropic and an ergogenic aid as it can enhance cognitive performance and improve strength and endurance; as such, it is classified as a psychoactive drug. All this means is that one must be wary of dosage limits, as it can cause some undesirable effects at higher doses. Caffeine may be found as different "variations" like caffeine anhydrous, dicaffeine malate, caffeine citrate, and pterostilbene-caffeine. It's the same molecule, just modified; anhydrous does not have any water molecules and may have more pronounced effects. Dicaffeine malate has some touted benefits; it can decrease fatigue with the additional *malic acid* molecule that's attached to caffeine, since *malic acid* plays a role in **ATP** synthesis. Caffeine citrate is caffeine attached to citric acid and can more readily be observed in the stomach. Pterostilbene-caffeine is caffeine attached to a molecule that's a derivative of **Resveratrol** and seems to enhance caffeine's effect on the body and has similar effects on the body to resveratrol.

Caffeine's main mechanism of action in the body is that it directly affects/blocks the actions of *adenosine,* a neuromodulator that's linked to drowsiness, arousal suppression, and sedation. It is also able to fit into the *adenosine* receptors, more so the A1 receptor which is responsible for sleep, without activating it thereby inhibiting *adenosine* activation. This promotes a wakefulness/alert effect.

- For those with caffeine tolerances, tolerance and withdrawal vary; no fixed month-long reset guarantees full effects. If a tolerance is already built, effects may not occur regardless of the dose amount.

Caffeine can also increase *serotonin* and *dopamine* levels in the brain, allowing for enhanced cognitive function, along with the synergistic effect of *analgesia (pain killer); this* can enhance performance due to a reduced perception of fatigue. It is believed to accomplish this from the antagonism of the *adenosine* receptors since the activation of the *Adenosine 1* and *Adenosine 2* receptor has been linked with decreased extracellular serotonin levels.

Benefits

- Enhance cognitive function
- Promote fat loss
- Strength increase

Side Effects

- High blood pressure
- Sleeplessness
- Anxiety
- Increased stress
- Nausea
- Tachycardia

Dosage

Count caffeine from all sources. FDA cites 400 mg/day as generally tolerated by most healthy adults; sensitivity, pregnancy, medicines and concentrated products change risk.

WARNING

Seek medical advice before use. Caffeine may affect **Diabetics** and those with heart conditions as caffeine has been seen to affect blood glucose levels and abnormal heart rhythms. Caffeine may affect those with **stomach disorders** since caffeine can increase stomach acid production. Caffeine affects serotonin levels, aggravating **bipolar** and **anxiety disorders** as well as **schizophrenia** and **Parkinson's disease**. Caffeine can also increase the amount of calcium that is released in urine which can negatively affect those with **osteoporosis.**

Carnosine *(Beta-alanyl-L-histidine)*

Carnosine exists in high concentrations within muscle tissues, especially during intervals involving force and high intensity for prolonged periods of time. Carnosine may in fact be more potent than creatine and act as a general antioxidant, but it is important to note a few key important factors before blowing your money. Carnosine supplementation WILL NOT increase carnosine concentrations in your muscles. Let me rephrase that. As carnosine supplementation increases, carnosine will enter muscles to an extent. However, the carnosine concentrations that can and will be achieved through this supplementation are insignificant and will not promote any benefits or effects.

So how do you get carnosine levels to rise? As you can tell by the name, carnosine is made up of **Beta-alanine** and L-Histidine. This is key to increasing carnosine levels.

- If you go back to **Beta-Alanine, it** is a bit more in-depth, but here's the simple answer. You need to supplement your diet with beta-alanine and l-histidine, carnosine's two substrates. These two amino acids are able to enter muscle cells at a significant rate, as this is where the synthesis of carnosine happens.

Beta-alanine is the rate-limiting producer of carnosine, meaning no beta-alanine, no carnosine. Although histidine is important, as a histidine deficiency occurs, so does a carnosine deficiency. It is important to supplement these two for optimal production. Hence the name CarnoSyn® is synonymous with carnosine increases in muscle tissues; however, it does not contain carnosine but beta-alanine.

Carnosine supplementation may be beneficial for older individuals as it has been considered a neurological peptide and may suppress *telomere (genetic material that protects chromosomes at the ends)* shortening that naturally progresses as we age. In the brain, carnosine is degraded by the enzymes *cellular carnosinase* and *serum carnosinase*; their activity seems to increase as we age. Although the full mechanism is not known, higher carnosine levels in the brain have been linked to elated mood and cognitive performance.

- Carnosine seems to slow *telomere* shortening rate and promote the production of cells with more "youthful" characteristics when compared to that of a younger person. The mechanism isn't fully known but may be related to *telomerase*, the mechanism that replaces the little bits of DNA that are lost in cell division which can result in damaged DNA or fused chromosomes.

Benefits

- Increase anaerobic capacity
- May improve strength
- Fatigue reduction
- Promote lean muscle
- Cognitive enhancer
- Anti-aging

Possible Side Effects

- Paresthesia
- Flushing
- Dry mouth

Dosage

- Adhere to label instructions

WARNING

Please consult a medical professional prior to administering carnosine. There is concern that carnosine may lower blood pressure to unsafe levels in those with **blood pressure problems.** Although data is promising, lack of research and data remain inconclusive for long-term use.

Cat's Claw *(Uncaria tomentosa)*

Cat's claw is a woody vine that can be found in the Amazon and some parts of Central America. Its name is derived from the description of its thorns that are shaped like a cat's claw. Traditionally it has been used to aid inflammatory and fatigue-related conditions. Currently, it can be seen marketed as an athletic supplement that is used to promote cartilage repair and as a possible estrogen inhibitor.

Prior to moving forward, it is important to note that there are other plants in the amazon that go by the same name that are not within the same genus and are used for the same reasons; however, they may affect the body differently, so it is advised to acquire an extract that is standardized with the bioactive alkaloids.

The active components of this plant have been observed to be the *Pentacyclic Alkaloids: mitraphylline* and *isomitraphylline* as well as *quinic acid.*

- *Mitraphylline* has been seen to inhibit the inflammatory *cytokines (messengers) Interleukin (IL)-1α, IL-4, IL-6* and *IL-17* and *Tumor Necrosis Factor-α* which can lead to reduced inflammatory response rather than chronic ones, resulting in reducing stress to the localized area. High inflammatory levels can result in oxidative damage from *Reactive Oxygen Species (Oxidant)* produced from *macrophages* and *granulocytes (white blood cells).*
- This can be beneficial for inflammatory conditions such as *Osteo/Rheumatoid* arthritis; additionally, it has been observed to promote and maintain *IGF-1* levels in *chondrocytes (involved in the creation and maintenance of cartilage)* as inflammatory *cytokines* can reduce *IGF-1* synthesis. *IGF-1* is involved in the synthesis of *proteoglycans (fibrous proteins that make up cartilage)* via *chondrocytes.* When combined with **Maca** it seems to further increase *IGF-1.*
- *Quinic Acid* seems to have an antioxidant effect, as it was found to protect under oxidative and heat stress for longer periods of time. This may be due to the extract's ability to maintain/preserve *Catalase (antioxidant)* levels. High antioxidant levels can promote DNA repair in damaged cells.

It has also been seen to be anti-estrogenic, although the full mechanism is not fully known; it seems to block estrogen receptor activation by inhibiting the synthesis of signaling molecules required for activation/signaling.

Benefits

- Anti-inflammatory
- Cartilage repair
- Antioxidant
- Promote cell repair
- Testosterone booster

Possible Side Effects

- Nausea
- Dizziness
- Diarrhea

Dosage

- 250-350 mg a day of a standardized *Pentacyclic Oxindole Alkaloid* extract

WARNING

Please consult a medical professional prior to administering Cat's Claw. There is a concern that Cat's Claw may affect those diagnosed with **hormone-sensitive conditions** such as **breast cancer** or **endometriosis** as it can reduce estrogen levels in the body. It can worsen those with **auto-immune diseases** as Cat's Claw can make the immune system more active. It can also reduce clotting factors affecting those with **bleeding disorders.**

Cayenne (*Capsaicin*)

Cayenne is a pepper from the *Capsicum annuum* sub-family and is used to flavor dishes and can also be found in powder form. In cayenne, the actual bioactive compound that gives the chili its heat and spice is called *Capsaicin* and can be currently found marketed as a weight loss product and an anti-inflammatory. In some products Cayenne or *Capsaicin* can be referred to as *Hot Pepper Extract.*

Capsaicin acts via the adrenaline receptors, specifically *Transient Receptor Potential Cation Channels Subfamily Vanilloid (TRPV).* These receptors are known to modulate/interact with *vanilloid compounds* which *Capsaicinoids* are a class of *vanilloids* — making *capsaicin* an active *TRPV1 agonist (activator).*

- *TRPV1* is involved in the regulation of body heat and the sensation of heat, as it is sensitive and responsive to the heat that is experienced. *Capsaicin* can lower the *TRPV1* threshold of heat needed to be activated all the way down to current room temperature. **Arachidonic Acid (AA)** is also a known *agonist*; it is believed that one can experience this effect through symptoms from a cold as AA is released from white blood cells.
    - *TRPV1* activation causes a calcium influx; calcium influxes are strong cellular signalers. Cellular calcium influxes may be the reason for the athletic improvements observed.

PGC-1α has been observed to be increased through *TRPV1* activation. *PGC-1α* is the master regulator of *mitochondrial biogenesis; increasing* mitochondrial mass in a cell can increase endurance. *PGC-1α* has also been seen to promote the secretion of *β-aminoisobutyric acid* which has been linked to thermogenic gene activation in white fat which can result in additional fat loss. Enhanced endurance may also be increased due to the use of fatty acids as fuel resulting from *adrenergic* activity in the release of *catecholamines (adrenaline, noradrenaline, dopamine)* down cascade of *TRPV1* activation.

TRPV1 stimulation can increase pain sensation in neuropathic pain since TRPV1 modulates body heat and pain. It is recommended to supplement with creatine as well to reduce the possibility of calcium-induced cytotoxicity (cell death) due to the calcium influx caused by TRPV1 activation.

- Anti-inflammatory and *analgesic* effects seem to come from *Capsaicin* inhibiting *IkB (Reference* ***Boswellia serrata*** *on IkB for a deeper explanation).*

Benefits

- Aid fat loss
- Thermogenic
- Enhance endurance
- Promote muscle growth

Side Effects

- Gastrointestinal irritation
- Sweating
- Flushing
- Runny nose

Dosage

If using an extract, as always follow instruction labels diligently.

WARNING

Please consult a medical professional prior to administering Cayenne extract. There is concern that Cayenne extract may affect those diagnosed with conditions such as **Irritable Bowel Syndrome**, **ulcers**, **heartburn,** or **gastritis** as cayenne can cause irritation of the intestinal lining. Overconsuming cayenne extract can increase the risk of developing **ulcers**

Chia *(Salvia hispanica)*

Chia, or Chia seeds, have been generally used as a nutritional supplement with many proposed health benefits. Chia is a plant from the *Lamiaceae* family where mint, lavender and basil can be found; it's also native to the western-southwestern areas of the Americas. Chia Seeds are the edible part of the Chia plant and have been found to have a high omega fatty acid content; 60% of it is *Alpha-linolenic acid* (an **omega-3** fatty acid) and about 20% of it is *linoleic acid* (an **omega-6** fatty acid) along with the antioxidant *myricetin* (*Linoleic acid* can form GLA; conversion of *Alpha-Linolenic Acid* to EPA and *DHA* is limited, reference those pages for an in-depth explanation of benefits.). There is very little observed data on the health benefits that are special only to Chia seed consumption; since omega fatty acid, fiber, and antioxidant benefits can be obtained from other plants.

Chia seeds are high in protein containing a complete amino acid profile with all 9 essential amino acids; the seeds contain anywhere from 20% to 30% protein content. It has also been noted to be a suitable oil substitute.

Nutrition Content		100 g of Chia Seeds	Nutrition Content		100 g of Chia Seeds
Aspartic Acid	-	1.69 g	Theonine	-	710 mg
Serine	-	1.05 g	Glutamic acid	-	3.5 g
Glycine	-	950 mg	Alanine	-	1.05 g
Valine	-	950 mg	Cysteine	-	410 mg
Methionine	-	590 mg	Isoleucine	-	800 mg
Leucine	-	1.37 g	Tryptophan	-	440 mg
Tyrosine	-	560 mg	Phenylalanine	-	1.01 g
Lysine	-	970 mg	Histidine	-	530 mg
Arginine	-	2.14 g	Proline	-	770 mg

Chia seeds have not been shown to reduce or suppress appetite; blood pressure seems to be unaffected as well. Although topical applications such as oil/lotion on the skin seem to enhance physical appearance.

- This may be due to the antioxidant *myricetin* as it has been shown to have positive benefits against *lipid peroxidation* and *oxidation*, along with the benefits from the fatty acid content.

Proposed benefits (evidence varies)

- Enhance skin appearance
- Cooking oil substitute
- Antioxidant
- Good source of omega fatty acids

Side Effects

- **Data is still inconclusive no common side effects have been found**

Dosage

- 25-100 grams for daily fiber, protein, and omega acids consumption; there is no set optimal dose.

WARNING

Seek medical advice before use. Do not swallow dry seeds without fluid because they swell. Chia's ALA is omega-3, and ordinary chia consumption is not established to increase **prostate-cancer** risk.

Chitosan

Chitosan is a *polysaccharide (complex carbohydrate)* that comes from the shells of shrimps. One thing to note is that humans cannot digest chitosan. However, it is being looked upon as a possible carrier of pharmaceuticals as it can be made into a bio-polymer. Outside of its biomedical uses, it has been commercially used as biodegradable products such as pesticides and polyurethane paint. In humans, it seems to have an interesting property. When digested with a meal, chitosan has the ability to bind to fats and cholesterol before they have a chance to be absorbed by the body. This data, however, has been found to be inconclusive as various studies have shown that the effect is very prominent while others have shown the exact opposite. Interaction with the body has shown the same variability; further, more concise testing needs to be done to assess any beneficial value. Human studies do not establish substantial weight loss; binding fat in a model does not guarantee a useful clinical effect.

Chitosan's mechanism of action seems to be derived from the fact that chitosan is a *lipophilic (fat-loving)* molecule. It has been seen to inhibit oil absorption in the small intestines by attaching itself to fat molecules; this keeps them from entering the intestinal wall since chitosan cannot be absorbed by the body, thereby carrying the fat molecules out of the body. Once again, the results of this mechanism have had mixed results.

- It is believed that chitosan has positively charged ions that are attracted to fat molecules allowing them to discharge. This, theoretically, can trap fat molecules, making them indigestible, limiting the amount of fat that is actually absorbed by the body.

Chitosan has also been seen to positively interact with inflammatory *cytokines (signalers) Interleukin (IL)-1, IL-2,* and *Tumor Necrosis Factor-α.* However, elsewhere chitosan has been seen to increase *IL-1* and *IL-6* pro-inflammatory reaction while reducing the inflammatory reaction of *Tumor Necrosis Factor-α;* this can augment chronic inflammatory conditions in those who are obese — increasing the cardiometabolic risk, down-cascade of either oxidative stress caused by chronic inflammation or the inflammation alone.

Proposed benefits (evidence varies)

- May promote fat loss
- May lower cholesterol

Side Effects

- Gas
- Nausea
- Constipation
- Upset stomach

Dosage

There is no set serving amount or dosage; always follow instruction labels diligently.

- In clinical studies, 2.4 grams a day was used for weight loss.

WARNING

Seek medical advice before use. Chitosan may affect people with a **shellfish allergy** as chitosan is derived from shellfish products. Those with **diabetes** or **obesity-related complications**, chitosan could negatively affect or exacerbate conditions or symptoms. Chitosan may also inhibit the absorption of fat-soluble vitamins like Vitamin A, D, E, and K.

Chlorella

Chlorella is a single-celled green alga from the *Chlorellaceae* family. *Chlorella pyrenoidosa* or *Chlorella vulgaris* are the most commonly found Chlorella supplements in the supplement market, although there are various other species of Chlorella out there. However, nutrient content and active **Vitamin B12** vary by species and **product; universal benefits are not established**, and like other algae, it has been observed to be a good *chelator (bonds to metals to assist the body in removal)* of heavy metals and a good source of *chlorophyll.* A commonly found form of Chlorella is Chlorella with a "Broken Cell Wall" which is touted to enhance detoxification, although there currently is no concrete data to support its efficacy in comparison to regular Chlorella.

The metal chelating activity of Chlorella allows the scavenging of toxic elements in the human gut having a high affinity towards arsenic and mercury. Chelation can become a double-edged sword; however, as the scavenging may also lead to nutritional deficiencies of essential trace metals. Since Chlorella is a good source of iron, Vitamin B12 supports red-cell formation but is not an established enhancer of iron absorption; this can help increase *hemoglobin (a protein that helps transport oxygen in blood cells)* levels.

Chlorophyll is the pigment plants use to capture light energy. It shares a ring framework with heme, but contains magnesium rather than heme's iron and is not an oxygen-carrying protein. Laboratory studies of chlorophyll derivatives explore carcinogen binding, oxidative signaling and cancer-cell growth. These findings do not mean that consumed chlorophyll replaces *hemoglobin*, oxygenates tumors or treats cancer. Cancer cells can also use oxygen; their altered metabolism is more complex than an exclusively anaerobic state. Clinical anticancer benefit from th*is supplement is unestablished.*

- *Chlorophyll* found in Chlorella has also been seen to reduce *dioxin (toxic environmental pollutants that can cause cancer and various other diseases)* absorption in the body; it is believed that this happens because *chlorophyllin (derivative of chlorophyll)* can form complexes with some *dioxins* reducing their absorption and biotoxicity, although currently, the mechanism by which it accomplishes this is unclear.

Benefits

- Vegan source of nutrients
- Heavy metal chelator
- Antioxidant
- Cardiovascular health aid
- Anti-cancerous

Side Effects

- Green stool
- Upset stomach
- Diarrhea
- Light sensitivity
- Gas
- Stomach cramps

Dosage

- Recommended dosage in studies is 6-10 grams daily

WARNING

Seek medical advice before use. Chlorella may affect people with **autoimmune diseases** such as **lupus,** and **rheumatoid arthritis** as chlorella can stimulate the immune system. If you're **allergic** to **molds,** you may have an allergic reaction to chlorella.

Choline (AKA Choline Bitartrate)

Choline is seen as a vitamin-like substance that is essential in the body as it is involved in motor functions in the form of *acetylcholine*. *Acetylcholine* is a neurotransmitter that is involved in memory and muscle control. In muscle control, acetylcholine is what allows your muscles to contract through interaction with *Neuromuscular Junctions*; these are junctions on your muscle tissue where acetylcholine is used to "excite" an *axon terminal* allowing the muscle to contract. Choline is essential to normal cognitive function, although supplementation has not established general cognitive enhancement as *choline acetyltransferase* converts choline into *acetylcholine* increasing neurotransmitter levels; this can then enhance *nicotinic receptors* that are involved in memory function.

Choline is a huge source of *methyl groups* that are necessary for the metabolism of various other essential molecules in the body like the synthesis of lipids and body detoxification. Choline is also needed to create **lecithin** (ref page for further benefits) and is a major source of methyl groups needed to create **SAMe** (Ref page for further benefits). Choline plays a huge role in the body's functions outside of *acetylcholine* synthesis; it is involved in cell signaling and homocysteine balance/reduction.

Choline can be oxidized by mitochondria in the body to make betaine; Choline can then *vicariously (actions produced through another molecule)* have cardioprotective properties as betaine has a high affinity for cardiac tissue. This is mainly due to its ability to maintain healthy homocysteine levels. Betaine is able to donate a methyl group to homocysteine, converting it into **L-Methionine**.

- High levels of homocysteine have been implicated in the activation of the enzyme *elastase (breaks down elastic fibers known as elastin)*. This can compromise and degenerate the integrity of arterial tissue/membrane. Homocysteine can additionally induce collagen production in smooth muscle cells leading to the creation of *fibrosis* and plaque build-up within arterial walls. This causes homocysteine to further activate *cyclins*; *cyclin* is a protein involved in cell *proliferation (cell multiplication)* increasing the risk of heart disease.

Betaine's conversion of homocysteine to methionine can promote creatine synthesis in the body as the added provision of methionine is able to produce **S-adenosyl methionine**, which is needed for creatine synthesis.

Proposed benefits (evidence varies)

- Enhanced cognitive function
- Reduced fatigue

Side Effects

- Sweating
- Diarrhea
- Fishy body odor
- Vomiting
- Upset stomach

Dosage

- Adult adequate intake is 425 mg/day for women and 550 mg/day for men from all sources; life-stage needs differ.
- The adult upper limit is 3,500 mg/day. Essentiality does not establish added sports or cognitive benefit from extra choline.

WARNING

Seek medical advice before use. Choline may increase the risk of **colon** and **rectum cancer**; however, choline has shown anti-cancerous effects along with Betaine and it is believed that the risk of colon cancer may be due to consuming high levels of choline.

Chondroitin Sulfate

Chondroitin is probably well known to those who suffer from joint degenerative diseases. Chondroitin is a *glycosaminoglycan (GAG)*; this molecule is vastly important in the makeup of joint cartilage. Chondroitin has been mainly used to help with joint pain and reduce the symptoms of arthritis. Generally, you will see chondroitin combined with **glucosamine**, although clinical results are inconsistent and depend on preparation and indication.

The *extracellular matrix (ECM)* that makes up joint cartilage is made up of **collagen** and *proteoglycans. Proteoglycans* are made up of a protein and a *GAG. Chondrocytes (involved in the creation and maintenance of cartilage)* are thinly spread throughout the *ECM.* This is important as *chondrocytes* synthesize the *proteoglycan* fibrous proteins that make up cartilage. Collagen fibers make up the structure and framework of the cartilage while a combination of *proteoglycans* attaches to the **Hyaluronan** (another *GAG*) backbone/structure to fill in the space of the collagen framework.

- *GAG*s are negatively charged; this allows them to attract water; this can also promote healthy *synovial* cells that can ultimately lead to proper *synovial fluid (fluid made from hyaluronan that lubricates joint movement)* production.

When *synovial* cells and chondrocytes become inflamed through years of damage, improper use, or as a result of improper health, this can cause a cycle that starts to build up from the previous cause. Inflamed *synovial* cells and *chondrocytes* lead to degraded *proteoglycans* resulting in a degraded collagen structure; this can also trigger inflammatory *cytokines (messengers)* that cause further pain and inflammation. This then results in an augmented synovial inflammation which can result in *chondrocyte apoptosis (programmed cell death).* The entire network then slowly starts to fall apart as each part can no longer support the function of the other, resulting in cartilage loss.

In theory, it is believed that increased chondroitin circulating levels in the body can then reduce connective tissue degradation and create an anabolic cartilage environment. This may be due to the *GAG* increase, which could lead to less inflammation and pain, which could ultimately lead to increased mobility.

Proposed benefits (evidence varies)

- May reduce joint pain and inflammation
- Proposed injury recovery aid
- May improve mobility
- May aid joint health

Possible Side Effects

- Upset stomach
- Leg swelling
- Headaches
- Eye lid swelling
- Hair loss
- Skin rash

Dosage

- The recommended dosage is about 1,000-1,200 mg broken into 2 dosages a day taken with food.

WARNING

Seek medical advice before use. Chondroitin may affect people with a previous **prostate cancer** condition as chondroitin may cause a reoccurrence. Chondroitin may make **asthma** symptoms worse, as well as decrease blood clotting in those with pre-existing **blood clotting disorders.**

Chromium *(Chromium Picolinate)*

Chromium, an essential mineral, is also an element that can be found on the Periodic Table sitting at atomic number 24. It is a Transition Metal where Manganese and Iron are found, and it's a shiny silver grey-ish metal. Chromium can most commonly be seen as Chromium Picolinate, but can also be found as Chromium chloride, Chromium citrate, Chromium dinicocysteinate, and Chromium nicotinate. Chromium has been seen to regulate glucose metabolism and insulin sensitivity in the body.

Chromium seems to react mainly through *Apo-chromodulin*. *Apo-chromodulin* is found inside cells; when *Apo-chromodulin* is loaded with Chromium Ions (can be loaded with a maximum of 4 ions total), it is then able to bind and activate the insulin receptor. More Chromium ions mean an amplified Insulin Receptor activation. This mechanism can then lead to improved glucose absorption at the cellular level. <u>However, high amounts of Chromium can also cause DNA damage.</u> Once insulin levels drop, insulin receptors no longer need to be sensitized, thereby releasing *chromodulin* as a whole, usually through urine.

- The increase in insulin response may also promote muscle growth in athletes as insulin has *mitogenic (cell division/multiplication)* functions as it can promote muscle *differentiation* and the *glucose* uptake into muscle cells can further enhance muscle growth. This can also aid fat loss as *glucose* will have a higher affinity to be used in muscle tissue after a workout rather than being stored in fat cells.

It has been observed that many diabetics may have chromium deficiencies, as high sugar diets can reduce chromium levels in the body, which would theoretically result in high insulin resistance. However, current studies have shown that it may not be the actual cause of diabetes, as there is no conclusive definition to state or test chromium deficiency. Some trials report glucose changes, but chromium is not an established treatment or cure for diabetes or a reliable bodybuilding aid.

<u>Proposed benefits (evidence varies)</u>

- May promote fat loss
- May enhance muscle growth
- Improved energy
- Antidiabetic

<u>Side Effects</u>

- Impaired coordination
- Headaches
- Mood swings
- Nausea
- Skin irritation

<u>Dosage</u>

- 300-500 mcg 2x alongside a meal, especially one rich in carbohydrates; high-dose safety is not established and kidney or liver injury has been reported (Reference warning below)

WARNING

Seek medical advice before use. Chromium might affect psychiatric conditions such as **depression, anxiety,** or **schizophrenia** as chromium can interact with serotonin. Please monitor any change in symptoms. Chromium can also affect those with **chromate** or **leather contact** allergies. Those diagnosed with **Diabetes** should consult with a doctor as chromium might affect insulin levels; medications may need to be modified. Chromium may cause kidney and liver damage with those suffering from **kidney** or **liver disease.**

Chrysin *(5,7- dihydroxyflavone)*

Chrysin, also referred to as *Passiflora caerulea* or honey extracts, is a *flavonoid* extracted from bee pollen, honey and a couple of types of Passionflowers *(Passiflora).* Chrysin can also be found in chamomile and some mushrooms. It can be currently seen marketed as a testosterone booster, although through oral consumption, bioavailability seems to be rather poor, as absorption into the body has shown extremely low metabolites in the blood to significantly boost testosterone. Animal and cell findings do not establish an effective or safe testosterone-boosting regimen in humans.

In preclinical work, chrysin has been investigated for effects on testosterone by inhibiting the *aromatization (conversion)* of testosterone into estrogen. Although data states poor absorption of chrysin, human studies do not establish these effects from oral supplementation. Chrysin sensitizes the testicles into creating more testosterone, by inhibiting *DAX-1 protein; DAX-1* protein negatively affects testosterone production as it is a negative regulator of *Steroidogenic Factor 1 (SF-1)* and *Steroidogenic Acute Regulatory (StAR)* protein. *SF-1* regulates testosterone synthesis in *Leydig (located in the testes/ovaries)* cells while *StAR (primarily located in steroid producing cells)* is a protein that regulates mitochondrial *cholesterol transfer; cholesterol* is the basis for steroid hormone production.

If oral absorption methods can be improved, then chrysin could have an enhanced benefit on testosterone synthesis.

In laboratory models, Chrysin has also shown activity against *colon cancer cells* as it has been seen to promote *apoptosis (programmed cell death)* in cancer cells. It has been seen that the mechanism of this effect was through activating and increasing *caspase-3* and *-9* as well as *BAX* protein activity. *Caspase-3* is involved in executing *apoptosis* while *caspase-9* initiates it; these are enzymes that are crucial in the *apoptotic* pathway. *BAX* protein is involved and can activate *apoptosis* in a cell.

Proposed benefits (evidence varies)

- May promote testosterone production
- Anti-cancerous

Possible Side Effects

- **Not enough is known, or no side effects have been reported**

Dosage

- No established effective and safe human testosterone-boosting dose exists.
  - Experimental animal or injection doses should not be translated into self-treatment.

WARNING

Seek medical advice before use. Chrysin might increase bleeding or bruising in those with pre-existing **bleeding disorders**. Not enough is known about Chrysin; data is still early in its development; as such, no long term or known adverse effects have been noted or recorded. Please use under medical supervision.

Cinnamon Extract

There are various varieties of cinnamon; Cinnamomum *cassia* is the most commercially available cinnamon product. However, *Cinnamomum verum* has been seen to be of the highest quality and be considered "True Cinnamon." Cinnamon has been revered for its taste and flavor as a spice; however, currently it has been seen as a health supplement, marketed as antidiabetic since it has been shown to positively regulate glucose levels and absorption in people. Cinnamon does not have to be bought as a supplement; however, it may be beneficial to acquire as an extract since it does contain a liver toxin known as *coumarin* and coumarin content varies and extraction does not guarantee its removal. Brewing or sediment does not reliably remove *coumarin. Cinnamomum verum*, AKA Ceylon cinnamon, has been seen to contain lower amounts of *coumarin.*

- *Coumarin can become harmful at high doses and is considered a potential liver toxin; it is not a flavonoid, and cinnamaldehyde supplies much* of cinnamon's characteristic flavor.

The active components in cinnamon are called *methylhydroxychalcone polymers (MHCP). MHCP* has been seen to be a slower acting insulin-*mimetic (can act like)*, as it is able to activate insulin receptors on fat cells in the same fashion as insulin. This may enhance and promote insulin sensitivity and glucose uptake. This mechanism may amplify the body's own insulin response when taken with a carbohydrate-rich meal since insulin takes 10 minutes to have an effect on glucose. By doing this, *MHCP* takes about 30 minutes to an hour to have an effect on glucose.

- Since *MHCP* increases insulin sensitivity, theoretically, glucose should be transported effectively to the appropriate destination instead of floating around in the blood or stored in fat cells.

Cinnamon or cinnamon extract has been seen to inhibit the digestive enzymes *α-glucosidase (breaks down carbohydrates and starches into glucose)* and *sucrase (aids the breakdown of carbohydrates into glucose)*, thereby reducing the rate at which glucose is absorbed into the body which can result in lower blood glucose levels. Please note that a lot of research shows promise for diabetics; in theory, this should help in body composition but not to the extent that would be significant.

Proposed benefits (evidence varies)

- Antidiabetic
- Aid insulin sensitivity
- Lower glucose levels

Possible Side Effects

- Mouth or lip irritation
- Upset stomach

Dosage

- Trials use varied preparations; evidence *does not clearly support cinnamon for treating diabetes or weight loss.*

WARNING

Seek medical advice before use. **It is important to note that cinnamon contains Coumarin, a potential liver toxin, especially in cassia. Product content varies substantially.** Some people have a hard time metabolizing this chemical. The amount set for safety is 0.1 mg per 2.2 lbs of body weight. Ceylon cinnamon generally contains much less coumarin than cassia; fixed amounts cannot be assumed for all products. **Diabetics,** please consult with a doctor as cinnamon or cinnamon extract can affect glucose levels; insulin medication may need to be adjusted accordingly.

Cissus quadrangularis

Cissus quadrangularis is part of the grape family, *Vitaceae,* and has been used in traditional Ayurvedic and Siddha medicine as it was believed to help heal broken bones. The stems/branches are quadrangular shaped and can grow up to 4.9 ft; it is also native to India and parts of Africa. It can also be known as Ashti Shrinkhala, Devil's Backbone, and Veld Grape. Currently it's been marketed as a bone and joint aid as well as a menopause aid.

The main bioactive components of Cissus are *triterpenoids, ketosteroid*, and *friedelin.*

- While the mechanism isn't fully understood, it is believed that the joint pain reduction benefit that comes from Cissus seems to be the result of one of the *ketosteroids* found in Cissus, which is found to increase bone density, strength and calcium. This is mainly done through the increase of the *insulin-like growth factor (IGF-1)* that's specific to *osteoblasts (bone cell).*
- The *glycoside, 6'-O-trans-cinnamoyl-catalpol,* may positively affect *mesenchymal cells (cells that can become different types of cells ex. myocyte or osteoblasts) differentiation (transformation)* into *osteoblasts.* This may be the reason for the accelerated bone healing as increased levels of *runt-related transcription factor 2* protein *(the protein is needed for a mesenchymal cell to become an osteoblast)* have been seen following Cissus supplementation.
- *6'-O-trans-cinnamoyl-catalpol* may also promote bone growth as it is believed to activate *mitogen-activated protein kinases (involved in cell differentiation)* in relation to *osteoblast differentiation; it* is believed that this is done due to the increase in *alkaline phosphatase (a byproduct of osteoblast activity)* observed following Cissus supplementation.
- Cissus's active components have also been seen to act as a *glucocorticoid (causes osteoblast death) antagonist (blocker/inhibitor)* which could also contribute to maintaining bone mineralization.

This can be beneficial for women who may have problems or want to prevent bone degradation, especially in those who have *osteoporosis* or have gone through menopause.

Cissus may enhance **GABA (involved in relaxation and sedation)** signaling, so combining this with a pre-workout of sorts could negate both effects and cause irregular heart-beats.

Benefits

- Improved joint and bone health
- May reduce pain and inflammation

Possible Side Effects

- Headache
- Dry mouth
- Intestinal gas
- Diarrhea

Dosage

- For joint pain 3,200 mg a day, if it's standardized 2.5% ketosteroid extract 300-600 mg a day

WARNING

Please consult with a medical professional prior to administering Cissus Quadrangularis. There is concern that Cissus Quadrangularis may affect **Diabetics**, as Cissus can affect blood sugar levels, insulin medication might need to be modified.

Citrulline (AKA L-Citrulline or Citrulline Malate)

Citrulline is a naturally occurring amino acid that plays a vital role in the removal of ammonia, a toxic chemical, from the body via the *Urea Cycle*. Citrulline supplementation converts citrulline into arginine in the kidneys which leads to higher arginine levels in the blood than arginine supplementation alone.

Citrulline has a major role in both the *Urea (removal of ammonia) Cycle* and *Nitric Oxide Cycle*.

- In the *Urea Cycle,* **arginase** converts arginine to **ornithine** and releases urea. Ornithine combines with carbamoyl phosphate to form citrulline; ammonia supplies nitrogen through carbamoyl phosphate rather than acting as a cofactor. Citrulline combines with aspartate to form argininosuccinate, which yields arginine and fumarate. The cycle removes nitrogen as *urea,* which is far less toxic than ammonia. Arginine can also enter nitric oxide synthesis; supplement effects on performance remain inconsistent.

Supplementing Citrulline can thereby increase arginine levels in the body, meaning it could also indirectly aid with lowering ammonia levels due to the increase of arginine while also augmenting the *Nitric Oxide Cycle.*

- As ammonia builds up in the muscles, fatigue becomes more apparent, especially with increasing effort. Since citrulline is an intermediate (part) of the *Urea Cycle,* it is able to buffer and remove ammonia, allowing for improved performance.

In the *Nitric Oxide Cycle*, arginine is converted into citrulline through *Nitric Oxide Synthase (NOS)* enzymes resulting in *Nitric Oxide*. Citrulline can then be converted back to *argininosuccinate* then further converted to arginine via the enzyme *argininosuccinate lyase; this complements* both cycles beneficially.

- There are 3 types of NOS: *endothelial NOS (eNOS)* blood vasodilator, *neuronal NOS (nNOS)* nervous system/neurotransmitter development *and inducible (iNOS)* involved in immune response such as inflammation, all of which follow the same structure for *Nitric Oxide* creation.

This mechanism can enhance performance through vasodilation while reducing fatigue through the removal of ammonia; it can additionally aid those with Vascular problems such as Erectile dysfunction or poor blood pressure. Mainly because both of these cycles work synergistically with each other.

Benefits

- Modulate blood pressure
- Erectile Dysfunction aid
- Increased stamina
- Increased endurance
- Vasodilator
- Reduce fatigue

Possible Side Effects

- Diarrhea

Dosage

- 1 gram of Citrulline 3x a day for vascular health
- For performance enhancement 6-8 grams of Citrulline 30 mins prior to the session

WARNING

Seek medical advice before use. Citrulline may affect those taking nitrates for **heart disease** or **erectile dysfunction** drugs; the addition of citrulline may cause an unsafe drop in blood pressure. For those with Citrullinemia, the added Citrulline can affect conversion and removal of ammonia; refer to a medical professional before taking.

Coconut Oil *(Copra Oil)*

Coconut oil is a highly saturated fat; this makes it more stable against oxidation compared to other oils. It is derived and refined from the white meat found on the inside of coconuts; the end product is generally composed of *Lauric Acid* and ***Medium-chain Triglycerides (MCT)***. Just a little over half of the coconut oil is made up of medium-chain triglycerides between 50-65%. The body treats these differently compared to other types of saturated fats. When *MCTs* are broken down they are turned into ketones; the body has a higher affinity for using these *ketones* as fuel instead of having them stored as fat like most saturated fats. *A key important note is that most of the therapeutic and beneficial effects seen are due to the MCT content, including fat-loss. Coconut oil has various mixed results and studies, some even stating that it can potentially be unhealthy due to the high concentrations of saturated content (eating raw coconuts is healthy as it is not refined oil). Lauric-acid-rich coconut oil is not equivalent to purified C8/C10 MCT oil; clinical findings for one cannot be assumed for the other.*

- MCTs are absorbed and metabolized differently from long-chain fats, but still supply calories and can contribute to fat storage when energy intake exceeds expenditure.
- Pooled clinical trials show coconut oil raises LDL cholesterol compared with non-tropical vegetable oils, despite often raising HDL too.

A proposed increase in ketone production does not establish meaningful weight loss from coconut oil. There is no need to replace all dietary fats with it; replacing unsaturated oils can worsen LDL cholesterol. Any weight change depends on the overall diet, energy intake and activity. Trials of purified MCT oil cannot establish the same effects for whole coconut oil, whose fatty-acid profile differs.

On the external benefits of coconut oil, it has been shown to have more protective and hydrating effects on hair and skin than conditioners and lotions, while having extremely low skin irritations reported. Oil pulling (swishing coconut oil in the mouth) benefits may be due to the *lauric acid* found in coconut oil, but the proposed soap-forming reaction in saliva is not an established explanation; oil pulling should not replace brushing, flossing or dental care.

Proposed benefits (evidence varies)

- May aid fat loss
- Anti-bacterial
- Improve hair/skin
- Modulate cholesterol levels

Possible Side Effects

- May increase cholesterol (HDL and LDL proportions)

Dosage

- 1-2 tablespoons daily or in the replacement of other fats used in your diet

WARNING

Seek medical advice before use. Although research is fairly new, and data is either lacking, or no common side effects might not be apparent, coconut oil may affect you individually. Currently, there is inconclusive and contradictive data to support its health benefits.

Cod Liver Oil

Cod liver oil, like many other **fish oils** that can be found on the market, contains **omega-3** *polyunsaturated fatty acids* with the addition of **Vitamin A** and **Vitamin D** as well (reference Vitamin A and D for additional health benefits). Generally, it can be found marketed as a joint health aid, a *nootropic (cognitive aid)* and to reduce the risk of *Cardiovascular Disease.* Ordinary supplements do not reliably prevent *cardiovascular* events; selected prescription products have different evidence.

The omega-3 fatty acids found in Cod Liver Oil are *docosahexaenoic acid (DHA)* and *eicosapentaenoic acid (EPA); the* body makes limited amounts from essential ALA; dietary EPA and DHA are not equivalent to ALA. *DHA* has nootropic effects that are further enhanced by *EPA's* ability to increase blood flow to the brain; this may very well aid those diagnosed with cognitive problems. It is important to note, those who regularly DO NOT eat fish as part of their diet, *DHA* may increase cognitive functions, but for those that DO, benefits *may not be seen from this supplement*.

EPA's and DHA's anti-inflammatory effects may be due to the increase of *EPA* concentrations in *Neutrophils* following supplementation. *EPA* seems to *attenuate (slow/reduce) leukotriene B4 (inflammatory signaler)* production in *neutrophils (a type of white blood cell).* It is believed that it accomplishes this through the suppression of gene transcription for *phosphoinositide 3-kinase (PI3K)* activation. *PIK3* controls intra-cellular signaling and modulation of lipids like **arachidonic acid (AA)**. Modulating the production of *eicosanoids* can either inhibit or cause inflammation.

- *Leukotrienes* are a family of *eicosanoids; eicosanoids* are signaling molecules released from AA oxidation that are involved in inflammatory responses.

This anti-inflammatory mechanism has also been seen in the brain and may be the reason for its cognitive-enhancing effects, as pro-inflammatory *cytokines (messenger),* like *interleukin-1β,* have been seen in neural tissue of stressed and depressed individuals. Inflammation in neurons can decrease *neurogenesis (creation of nerves).* A secondary mechanism is believed through *$G_s$alpha subunit (a G Protein that activates and is involved in neurotransmission)* modification; this allows them to more effectively activate receptors on the cell. This *G Protein* has been seen to be an incomplete complex in depressed individuals.

Proposed benefits (evidence varies)

- Promotes joint health
- Improved hair & skin health
- May lower cholesterol
- Anti-inflammatory
- Nootropic
- Antidepressant

Possible Side Effects

- May reduce clotting abilities
- Heartburn
- Fishy burps
- Loose stools
- Nausea
- Bad breath

Dosage

- Amounts of EPA, DHA, Vitamin A and Vitamin D vary by product; count all sources. High-dose use requires medical guidance, not a routine six-dose regimen.

WARNING

Seek medical advice before use. Cod Liver Oil may drop your blood pressure to unsafe levels in those with **blood pressure problems**. Excess preformed Vitamin A can harm the liver and fetus; high-dose omega-3s can increase atrial-fibrillation risk.

Coenzyme Q10 (CoQ10)

Coenzyme Q10 cycles between oxidized **ubiquinone** and reduced **ubiquinol; a semiquinone intermediate also participates in electron transfer.** CoQ10 is the name for the compound, not a separate third supplement form. Cells interconvert these states as electrons move through mitochondrial reactions. Oral absorption depends on formulation, food and the individual. Ubiquinol may produce higher blood levels in some comparisons, but a universal advantage after age 40 and a fixed 95% whole-body proportion are not established. A biochemical role in energy production does not guarantee more energy or better athletic performance from extra supplementation.

CoQ10 is a fat-soluble vitamin-like substance found in any "breathing" cell that helps the body produce energy through **ATP**. The majority of the body's energy is created this way which is why you will find the highest amounts of CoQ10 in the heart, brain, liver, and kidneys. *Statins inhibit the Mevalonate Pathway, shared by cholesterol and CoQ10 synthesis; this does not establish routine relief of statin symptoms from supplements, and not all blood-pressure medicines block this pathway.*

- CoQ10 and its derivatives interact with the *electron transport chain* within the *mitochondria* to promote ATP production; this is where its energy-enhancing benefits derive from.

Supplementation of CoQ10 or its derivatives can provide protective benefit and proper function in the organs listed above. CoQ10 seems to protect the *endothelium (lining of the circulatory system)* from *LDL cholesterol* by inhibiting *NF-kB* derived inflammatory effects. Additionally, it is able to combat oxidative damage through *REDOX (loss or gain of an electron in a molecule)* interaction in practically every cell in the body while also maintaining *Nitric Oxide* activity through the same mechanism, leading to vasodilation aiding cardiovascular health along with providing athletic benefit for athletes.

- Low CoQ10 levels have been studied in fibromyalgia and heart disease, but association does not establish that deficiency causes these conditions. Supplementation of CoQ10 has been shown to increase concentrations in heart tissue, which can help reduce damage to the heart.

Proposed benefits (evidence varies)

- Cardiovascular health aid
- Possible antidepressant
- Enhance athleticism
- Promote energy
- Vasodilator
- Antioxidant

Possible Side Effects

- Nausea
- Loss of appetite
- Stomach discomfort
- Rash
- Diarrhea
- Dizziness

Dosage

- The recommended dosage is 90 mg a day with a meal for general health
- 200 mg prior to working out for older athletes looking to increase strength

WARNING

Seek medical advice before use. CoQ10 can affect those undergoing **chemotherapy**, as CoQ10 can reduce the effects of chemotherapy. CoQ10 can lower blood pressure to unsafe levels in those taking **blood pressure medication**. Review interactions with warfarin and insulin; clinical benefits remain indication-specific.

Collagen

Collagen is a structural protein that is in various connective tissues. The types that can be commonly found in the human body are as follows: skin, bone, and tendon are considered *Type I Collagen; the density* of mineralization decides the structure. Cartilage is considered *Type II Collagen. Reticular fiber (a mesh/network of fibers that make up soft tissues like the organs/liver)* is considered Type III Collagen. *Basal lamina (sits under the epithelium, lining the surfaces of organs)* is considered *Type IV Collagen. Type V Collagen* is generally found in the hair and placenta. The most commonly marketed Collagen is *Type II Collagen; its main* benefits are for joint relief and for those who suffer from arthritis.

Hydrolyzed collagen and undenatured collagen are some of the variations that can commonly be found in supplemental collagen, which differ in processing, dose and proposed mechanisms; their clinical findings are not interchangeable. It's important to note that collagen consumption is **NOT VEGAN-FRIENDLY** as it's derived from animals. Supplementation of *Type II collagen* can be beneficial at aiding joint inflammatory diseases.

- *Type-II Collagen* is collagen that is found and used to create cartilage on the surface of joints. Immune responses to cartilage components can contribute to rheumatoid arthritis, but arthritis is not simply caused by excess collagen release.

Undenatured *Type II Collagen* has been investigated for reducing joint symptoms as one would regularly see from the process mentioned above, by creating a type of "tolerance," to the level of *Type II Collagen* that creates this inflammatory response. It accomplishes this by affecting *dendritic cells. Dendritic cells* are the basis in response to *autoimmune inflammatory* diseases like rheumatoid arthritis as they are capable of producing immune responses such as inflammation.

- When one consumes something that contains *Type II Collagen,* the collagen interacts with *Peyer's patches* which are areas in the intestines with a high level of *dendritic cells.* With continued consumption, *dendritic cells* can change (become de-sensitized to collagen in the body in a sense) and induce a reduced response from T-Cells; a T-cell is a type of white blood cell (Elevated T-Cells levels have been found in the synovial fluid in those with rheumatoid arthritis). This changed response is then able to indirectly influence the production of *interleukin 10 (IL-10).* The release of *IL-10* inhibits the release of IL-17; IL-17 can drastically affect inflammatory diseases in a negative manner; this can attenuate inflammation in those with arthritis.

Elsewhere, hydrolyzed collagen has been said to benefit and support hair, nail, skin and bone health as it is higher in Type I and III Collagen. *(Refer.* ***Chondroitin*** *for synergistic benefits)* *A 2025 meta-analysis found skin benefits unconfirmed in high-quality or non-industry-funded trials; formulations and endpoints matter.*

Proposed benefits (evidence varies)

- Reduce joint inflammation
- Rheumatoid arthritis aid
- Reduce joint pain
- Improve skin and hair health

Possible Side Effects

Not enough is known or has been reported to asses common side effects

Dosage

- Undenatured collagen 40 mg/day has been studied for joint symptoms; this is not an established rheumatoid-arthritis treatment.

WARNING

Seek medical advice before use. Collagen derived from bovine (cow) as it might be contaminated and carry **Mad Cow** disease.

Colostrum

Colostrum can also be referred to as a type of bio-active peptide. Colostrum comes from the milk that's first produced after pregnancy by mammals to support newborns, providing them the antibodies, enzymes and growth factors needed in the otherwise undeveloped newborns. The main reason colostrum is so effective in newborn babies is that their intestines and intestinal walls are undeveloped; this allows the growth factors to easily enter the body; however, the majority of adults have fully developed intestines and intestinal walls that would otherwise break down the peptide into nothing more than whey protein or exert localized effects to the area and provide intestinal benefit. The most commonly found colostrum in the market is *bovine colostrum;* it also has been seen to have a high nutrient profile. It has even been found to have *IGF-1 (Insulin-like Growth Factor)* concentrations that are up to 10x more than that in milk, just as well as various *TGF (Transforming Growth Factor).*

- The problem is not the active content matrix of colostrum, but its ability to enter the body. Refer back to Bio-Active Peptides and the explanation of different types of peptides and their overall design and purpose and how their purpose is negated as they are broken down through the digestive tract. **All these growth factors have their own peptide design and are found in colostrum. The problem is that they will be broken down by the adult digestive tract prior to being absorbed in the intestinal tract.**

The only benefit observed in relation to performance was a reduction in intestinal inflammation due to prolonged high-intensity cardiovascular events. The effects seen are from the localized intestinal benefit achieved by colostrum reducing heat-induced intestinal permeability.

- An increase in intestinal permeability can allow "bad bacteria" the ability to cross the intestinal barrier thereby causing system-wide inflammatory and oxidative response as the breakdown of Gram-negative bacteria releases *Lipopolysaccharide (LPS)* an endotoxin that activates *Toll Like Receptor 4 (involved in inflammatory signals).* The increase in intestinal integrity may benefit those suffering from intestinal problems or who are prone to infections.

It is believed that Colostrum is able to provide intestinal support as it contains *Epidermal growth factor (EGF); this* can activate *EGF Receptors* in the localized area; this has been seen to support/protect intestinal integrity. Additionally, it is believed that *EGF Receptor* activation can promote expression of *Heat Shock Protein 70 (HSP70); HSP70* is a protein that protects cells from metabolic and oxidative stress; they are able to refold and stabilize mutated proteins and aid with DNA repair.

Benefits

- Promote intestinal health
- May lead to enhanced nutrient delivery

Possible Side Effects

- Nausea
- Vomiting

Dosage

- 20-60 grams per day for intestinal health

WARNING

Please consult with a medical professional prior to administering Colostrum. There is concern that Colostrum may cause an **allergic reaction** in those with **lactose intolerances** or cow product allergies. Colostrum may increase immune system activity; this may affect those with **autoimmune disorders**.

Cordyceps militaris

Cordyceps militaris is a type of fungus from the *Clavicipitiaceae* family, in which the *ergot alkaloids* needed to make LSD are found. Most of the fungi found in the *Clavicipitiaceae* are considered parasitic to other fungi or bugs and toxic to humans or animals; however, *Cordyceps* has been used in traditional Chinese Medicine as an anti-aging and life-enhancing fungus. Currently, it can be found marketed as the same. It is important to note that extracts vary in potency dependent on batch and manufacturer type.

One of the main bioactive components of cordyceps militaris is *Cordycepin (3'-deoxyadenosine),* which is considered a *nucleotide analogue (structurally similar)* of *adenosine.* (Refer to **ATP** for a deeper explanation of adenosine's importance in the body.) The main mechanism that is derived from *Cordyceps* is its ability to alter *adenosine* concentrations increasing energy conversion. It has also been seen to be an exercise mimetic as it is able to increase blood flow to not only skeletal muscle tissue, but to non-skeletal muscle tissue such as the liver; this can also enhance *lactate* removal from the body. The enhanced oxygen utilization and *lactic acid* removal can lead to less oxidative stress from exercise.

- During exercise, lactate becomes a byproduct of the *Anaerobic Glycolysis System.* This is the process where *glycogen (broken down to glucose)* is used for energy instead of oxygen, resulting in Lactic acid creation. Lactic acid turns *Pyruvate (this breakdown creates energy for muscles to use)* into *lactate* by the enzyme *lactate dehydrogenase; this* process releases *H^+ Ions.*

This process will continue until the amount of H+ Ions has created an environment where *Glycolysis* is inhibited, and muscle contractility is diminished; reducing the amount of *H^+ Ions* through efficient Lactic acid removal can increase endurance in muscle tissue.

Cordyceps can increase free testosterone in the body; however, *Cordyceps* has been known to suppress the **Forskolin** increase in testosterone, through the inhibition of *Cyclic Adenosine Monophosphate (cAMP)* stimulated *steroidogenesis (hormone creation).* Forskolin promotes testosterone production via *cAMP* induced *steroidogenesis. Cordyceps'* main mechanism for increasing testosterone secretions is through *Steroidogenic Acute Regulatory (StAR)* protein upregulation. *StAR* proteins, primarily located in steroid producing cells, are proteins that regulate *cholesterol* transport/transfer into the mitochondria (this is the rate-limiting step in *steroidogenesis); cholesterol* is the basis for steroid hormone production.

Benefits

- Enhanced oxygen utilization
- Testosterone booster
- Increased energy/endurance

Possible Side Effects

- Dry mouth
- Nausea

Dosage

- Dosages from 300-450 mg to 1-3 grams of cordyceps extract have been seen as safe

WARNING

Please consult with a medical professional prior to administering Cordyceps. Cordyceps might affect the immune system; for people diagnosed with **auto-immune diseases,** such as **Lupus** or **Rheumatoid arthritis, symptoms** may worsen. For those with Bleeding disorders, cordyceps might affect blood clotting and increase the risk of bleeding.

Creatine

The most discussed and used supplement in the fitness industry next to pre-workouts. Creatine is a molecule that can rapidly produce and support **ATP** production, enhancing athletic performance; studied doses have a good safety record in healthy adults; natural production alone does not establish safety. If you read supplement labels, it can become extremely confusing when you start seeing the variations of creatine, ex. *Creatine monohydrate, micronized creatine, creatine malate, creatine hydrochloride, etc.* Generally, as a rule of thumb, creatine monohydrate tends to be the cheaper alternative as well as having competitive benefit with other variations of creatine. Additionally, there are not enough studies to prove any other form is better currently except for independent studies. Creatine takes time to build up in your body, so the longer you take it, the more observed effects in performance and strength that can be seen. This is one of the most effective options for those on a tight budget.

Since there are so many benefits to creatine supplementation, it can and will take 20 pages or so to break it down like most supplements in this book. This is a summary of those effects.

- Cell hydration; creatine pulls water into the cell as it enters, which then promotes a cellular anabolic state; in return, *glycogen synthesis (needed for anaerobic events such as heavy lifting)* is stimulated. In short, muscle cell growth happens while increasing your state of energy.
- Creatine supplementation reduces the body's need to create its own; Creatine synthesis takes about half of the body's methyl donors, taking it away from other processes that might need it.
- Creatine intake might also be able to increase carnosine stores in cells. (Reference **carnosine** and **beta-alanine**)
- Creatine is also able to get through the blood-brain barrier as the brain sees fit. If there is a neural deficiency it takes creatine in; it also cannot be overloaded like in muscle tissues. Currently it is believed that creatine use is more towards neuroprotection instead of cognitive enhancement.
- Creatine seems to protect against *excitotoxicity* by maintaining ATP levels. If cells are depleted of ATP, membrane integrity is compromised thereby allowing an influx of calcium ions into the cell, which in turn leads to *apoptosis (programmed cell death); this* can lead to muscle loss.

This applies to any cell, not only muscle cells. The longer ATP reserves can be kept up, the longer a cell can maintain its integrity, leading to less cell death although human trials have not established prevention of neuro illnesses such as Huntington's and Alzheimer's, due to *excitotoxicity*.

Benefits

- Antioxidant
- Increase strength
- Muscle gain

Possible Side Effects

- Gastrointestinal discomfort
- Muscle cramping
- Diarrhea
- Water retention (high doses)

Dosage

- Common adult maintenance dosing is 3–5 g/day. Optional loading uses about 20 g/day in divided doses for 5–7 days, then maintenance.

WARNING

Prior to administering creatine, please consult with a medical professional. Those with **kidney disease** or relevant me**dicines** need individualized advice; creatine is not established to worsen diabetes routinely. If you have Bipolar Disease, creatine might also worsen symptoms.

<u>Curcumin *(Diferuloylmethane)* AKA Turmeric Extract</u>

Curcumin is not an herb, but the main active compound of Turmeric, and can also be found in other plants of the same family such as ginger, but at much lower amounts. Curcumin can have anti-inflammatory effects and be an antioxidant. Curcumin may also have some anti-cancerous properties, but since it has poor bioavailability, some formulations use *piperine (black pepper extract)* to enhance exposure, but this is not mandatory and can increase drug-interaction risk. Curcumin can be traced back for its benefits in traditional Ayurvedic medicine

Curcumin is a *diarylheptanoid* which is a natural *phenol.* In laboratory and selected clinical studies, it has shown anti-inflammatory effects which can be beneficial for those with *rheumatoid arthritis and osteoarthritis* via reducing *Interleukin 1β, Interleukin-6* and *vascular cell adhesion molecule 1* which negatively affects arthritis. Curcumin has the ability to be protective against *glutamate (released after physical activity/exercise)* and *corticosterone (stress)* induced cell death, leading to a decrease in cell death; this can provide neuroprotective factors. The two other active compounds are *demethoxycurcumin (DMC)* and *bisdemethoxycurcumin (BDMC).*

- Curcumin is able to suppress tumor cell signaling by enhancing confirmation of *AP-1 (involved in cell growth, proliferation/multiplication)* signals. This means that it can suppress signals that aren't quite right and could lead to a tumor creation. It can also enhance the liver's ability to metabolize foreign substances such as antibiotics and alcohol.
- *DMC* and *BDMC* have been seen to positively modulate *dopamine* levels, maintaining appropriate levels needed for appropriate directed motor function, a proposed mechanism that does not establish clinical benefit in *Parkinson's Disease,* as dopamine *depletion* has been implicated in improper motor function.

Curcumin has a significant increase in *nitric oxide* synthesis increasing vasodilation; it can also increase *glucose* uptake to skeletal muscles which could lead to higher levels of **ATP**. It has also been seen to provide antioxidant activity due to its *methoxy* ring while also enhancing *endogenous (inside the body)* antioxidant activity. <u>High doses of curcumin may, however, lead to anti-androgenic effects as well as higher estrogen levels.</u>

<u>Proposed benefits (evidence varies)</u>

- Increase nitric oxide synthesis
- Antioxidant
- Increased energy
- Aid with arthritis
- Aid Parkinson's Disease symptoms
- Anti-inflammatory

<u>Possible Side Effects</u>

- Gastrointestinal discomfort
- Nausea
- Dizziness

<u>Dosage</u>

- General dosage may be anywhere from 80-500 mg a day
    - **In any extract, some products enhance absorp**tion with *piperine, polyvinyl pyrrolidone or phosphatidylcholine.* Check ingredient labels.

<u>***WARNING***</u>

Seek medical advice before use. Liver injury has been reported with some turmeric/curcumin products, especially highly bioavailable formulations; it is not restricted to prolonged high doses. If you are diagnosed with **Gallbladder** problems, curcumin might increase bile production and make symptoms worse. For those with **low blood pressure**, curcumin may make blood pressure dangerously low. Curcumin might also slow blood clotting in those with **clotting problems**. Curcumin can affect those in **hormone therapy**, increasing estrogen levels.

Dandelion *(Taraxacum officinale)*

Although most people see dandelions as flowers or pesky weeds, the plant has a long history of being used as a vegetable as well. These little flowers are completely edible. 10 grams of the dried flower contains 10-15% of your daily potassium needs with almost no calories. This, however, does not mean to pull the dandelions growing in your yard, as they might have had pesticide exposure in the past and could potentially harm your health. Dandelions have been used in traditional Chinese Medicine and were seen as beneficial to health, but in comparison to other antioxidant herbs/plants it's on the lower end of the scale; some users report it to be a slower gentle way to detox the body.

One of the main active components of the dandelion is *Taraxasterol*. There are two main methods of extract; ethanol and water extract such as tea. Ethanol extract seems to have the highest antioxidant concentrations compared to other extraction methods. <u>Dandelions can carry heavy metal/minerals from the soil it was grown in; buy from reputable brands to reduce the risk of this.</u>

- Dandelion tea has been shown to increase enzyme activity in the liver, particularly the ones responsible for detoxifying the body, meaning it can aid the liver at detoxing the body. Specifically, *UDP-glucuronosyl transferase; this* enzyme directly affects glucuronidation; glucuronidation is involved in the metabolism/breakdown of drugs, *androgens, estrogens,* and various other compounds that contain a sugar-like bond.
  - <u>*It's important to note* that *this means it can reduce the effects of particular drugs.*</u>

Water extracts have been shown to inhibit the actions of enzyme *α-glucosidase; this* enzyme aids in the absorption of carbohydrate and glucose by breaking them down; this may be due to *Taraxacum officinale (TOP) 1* and *2 polysaccharides (a type of carbohydrate).*

- The entire extract seems to possess anti-inflammatory effects; this includes *Taraxasterol* and *TOP 1 and 2; all* the bioactive compounds seem to be able to inhibit various inflammatory responses, to include suppressing *macrophages (type of white blood cell)* inflammatory response as well as *neutrophil (type of white blood cell)* activation; this may be due to affecting ion crossing at the cell membrane that would otherwise activate them.
  - <u>*This can negatively affect those with a compromised or underactive immune system*</u>.

<u>Benefits</u>

- Possible antidiabetic
- Aid in liver detox
- Anti-inflammatory

<u>Possible Side Effects</u>

- Heartburn
- Diarrhea
- Upset stomach

<u>Dosage</u>

- The standardized dosage has not been set at this time; if using an extract, as always follow instruction labels diligently.

<u>***WARNING***</u>

Please consult with a medical professional prior to administering Dandelion supplements in your diet. Although there's a long history of dandelion use, clinical data and research are still new and/or lacking. There is concern that Dandelion supplements can cause an allergic reaction in individuals with **Ragweed** allergies. There is concern that Dandelion extracts can also affect those with underactive **Immune Systems**, as Dandelion extracts have been seen to reduce inflammatory/immune response.

Dang Quai *(Angelica gigas)*

DO NOT CONFUSE THIS HERB WITH DONG QUAI (reference **Dong Quai**) although similar in name, they are two different herbs used by two different groups of people for different health benefits. Confusing one with the other could lead to adverse health effects. Dang Quai, also known as Korean Angelica, refers to *Angelica gigas* that has been used in traditional Korean Medicine, generally to improve women's health, such as bone density, menstrual irregularities and symptoms as well as menopausal symptoms. Most proposed effects remain preclinical; species-specific human efficacy and dosing are not established.

Decursin and *decursinol angelate* seem to be the main active components, and they also appear to be *lipophilic (fat-soluble). There are different extract methods, each focused on specific benefits, ex. immune function; this is accomplished by focusing on extracting different areas of the plants; be sure you read product labels prior to administering.*

- *Decursin angelate* has also been seen to reduce inflammation due to excessive *Nitric Oxide (NO)* production since *NO* is involved in blood flow, especially in chronic inflammatory conditions. This is accomplished through activating *Heme oxygenase-1 (HO-1)* as it is a stress-induced regulator of *NO*, meaning it can positively regulate *NO* preventing oxidative inflammation.
- The two compounds have been seen to be resistant to *glutamate*-induced *neurotoxicity (neuronal death); HO-1* expression has also been seen in the brain; this can reduce oxidative inflammation in the brain through the exact same mechanism mentioned above.
- *Decursin* and *decursinol angelate* can act as natural *phytoestrogens,* thereby stimulating estrogen and progesterone receptors. They do this by *being analogues (something that's similar in structure)* of female hormones.
    - Preventing estrogen reduction, through the stimulation of estrogen hormone receptors in menopausal women, bone degeneration can be slowed by balancing bone resorption (the breakdown of bones). This means that bone density and mineral content can be maintained at higher levels through the stimulation of estrogen hormones by *decursin and decursinol angelate,* theoretically speaking, this could resist osteoporosis in women.

Proposed benefits (evidence varies)

- Aid menstrual symptoms
- May prevent osteoporosis
- Anti-inflammatory
- Reduce menopause symptoms
- Promote neural protection
- Immune booster

Possible Side Effects

- Photosensitivity
- Low blood pressure

Dosage

Not enough is known to have a standard dosage, adhere to label instructions.

WARNING

Seek medical advice before use. Dang Quai can affect those who are going through **hormone therapy** or have **hormone-sensitive conditions** such as **endometriosis, endometrial, breast, or ovarian cancer** as it can act as estrogen within the body. Dang Quai may negatively affect those with **immune system problems** as Dang Quai has been seen to increase the production of B cells (cells that produce antibodies). Please note that evidence is within preliminary stages; full benefits/side effects have not fully been discovered or reported.

D-Aspartic Acid (D-AA)

D-Aspartic Acid is a non-essential amino acid that can be seen marketed as a testosterone booster and a nootropic (cognitive enhancer). Something that should be reiterated is that although it can boost testosterone, it should be noted that it CAN raise testosterone levels TEMPORARILY for a week and dosages need to be exact, with levels later returning to baseline after that. Outside of that, testosterone levels will not increase in normal healthy men. High doses of D-AA can also lower testosterone levels. Only those suffering from infertility may see a prominent increase in testosterone levels. Controlled trials in trained men do not establish increased testosterone, strength or muscle from supplementation.

D-AA's mechanism for achieving this effect is that it is able to cross the blood-brain barrier and stimulate the release of hormones, such as *growth, follicle* and *luteinizing hormones (LH).* It can also build up in the *gonads* (reproductive organs) promoting the production of testosterone mainly by *Steroidogenic Acute Regulatory (StAR)* protein upregulation.

- *StAR* proteins, primarily located in steroid producing cells, are proteins that regulate *cholesterol* transport/transfer into the mitochondria (this is the rate-limiting step in *steroidogenesis); cholesterol* is the basis for steroid hormone production.
- It seems that D-AA can influence the release of *luteinizing hormone,* which directly induces testosterone synthesis. D-AA can accumulate in *Adenohypophysis (releases LH)* which increases the rate that hormones are secreted, leading to cascade effects eventually causing the release of the *luteinizing hormone* and *growth hormone.* At higher doses, however, oxidative damage can start to accumulate in the testicles leading to a reduction of testosterone and sperm defect.
- D-AA may have potential with females in enhancing and prolonging reproductive potential.

D-AA can be broken down to D-Aspartate, which has been seen to act as a neurotransmitter as D-Aspartate can stimulate neuronal transmission after *depolarization (mechanism of opening or closing channels involved in neurotransmission).* Additionally, D-Aspartate can be converted into *N-methyl-D-aspartate (NMDA)* through the addition of a methyl group thereby interacting with *NMDA* receptors. Since *NMDA* receptor activation is involved in memory and learning through *synaptic plasticity,* it is believed to improve cognitive function. However, overstimulation can lead to *excitotoxicity (nerve cell death by overstimulation).*

Proposed benefits (evidence varies)

- Increase in free testosterone
- May promote fertility
- May enhance cognitive function

Possible Side Effects

- Higher estrogen levels

Dosage

- Study recommended dosage is 2-3 grams a day

WARNING

Seek medical advice before use. D-Aspartic Acid can affect those on **hormone therapy** as D-Aspartic Acid can affect hormone production. It is important to note as of current, there is a lack of clinical studies and human research. This lack of data and research suggests unknown long-term effects or side effects. However, there is suspicion that prolonged use of D-AA can have negative feedback on testosterone after prolonged use. Cycling has not been shown to prevent adverse hormonal effects or make the supplement effective.

Dehydroepiandrosterone

Dehydroepiandrosterone (DHEA), also called androstenolone or prasterone, is a steroid hormone that is abundantly found in the body. It is created by *the adrenal glands (kidneys), gonads (testis/ovary),* and the brain. Its primary function is being a precursor to the synthesis of testosterone and estrogen. DHEA can positively modulate hormones which can be beneficial to both men and women. In an adult male/female, DHEA peaks at puberty to 25 years of age and starts declining between 30-40 years of age.

- DHEA may have the ability to balance out *cortisol (stress hormone)* being that *cortisol* and DHEA are both induced by the *adrenocorticotropic hormone* and are always kept in ratio with each other. *Cortisol* & DHEA also use the same enzyme *cortisone reductase*; *cortisone* converts into *cortisol* and DHEA converts into **7-Keto DHEA; this** can aid *hypercortisolemia (high blood cortisol)* seen in those with schizophrenia, depression, anxiety, and bipolar disorder as the additional DHEA would compete with *cortisone,* reducing the amount of *cortisol* produced.
- The conversion from DHEA to **7-Keto-DHEA** by the enzyme *cortisone* is irreversible as this pathway is not multi-directional. This means that it cannot turn back into DHEA and since DHEA is the intermediate and basis for androgenic and estrogenic hormones, 7-Keto-DHEA cannot be further converted into estrogens; this can promote higher testosterone levels.
- DHEA may aid fat loss as it has been seen to reduce *Peroxisome proliferator-activated receptor gamma (PPARγ)* genetic protein needed to bind and store fat; PPARγ has also been seen to stimulate *uncoupling protein (involved in thermogenic/fat burning)* genetic expression in fat cells.
- DHEA can promote the increase of *IGF-1 (involved in various bodily functions)* in older individuals, which in turn has been seen to promote *osteocalcin (bone-building hormone)* expression in *osteoblasts (bone-forming cells); this* can promote bone mineralization (on selective areas), especially in women with *osteoporosis.*

Topical application will enhance skin appearance, but will also have near the exact same result as orally consumed DHEA. DHEA may be converted into estrogen instead and cause negative feedback if you have high testosterone levels.

Benefits

- Increase in testosterone
- Increase bone density
- May lower stress/anxiety
- Enhance skin appearance
- May help fat loss
- May reduce aging

Possible Side Effects

- Headache
- Fatigue
- High blood pressure
- Acne
- Mood changes
- Abnormal periods
- Facial hair growth
- Gynecomastia
- Insomnia

Dosage (Depending on standardized DHEA content dosages may vary)

- Over 40 years of age, 25-50 mg a day, for testosterone increase is 100-200 mg a day
- For skin health, 4 grams of topical cream at 10% DHEA content

WARNING

Please consult with a medical professional prior to administering DHEA. There is concern that DHEA may affect those with diabetes**, polycystic ovary syndrome**, **high cholesterol**, **thyroid**, **liver**, and **heart problems** at higher doses as it can interact with the hormones that regulate these functions. Those on **hormone therapy**, the same concerns apply as it may also counteract the therapy.

Deer Antler Velvet Extract

Deer Antler Velvet comes majorly from New Zealand and is harvested around June or December depending on the location. The antler is cut off around the base of the deer or elk to acquire the antler velvet while it is still in its pre-calcified state of growth, usually around 55 to 65 days of growth. Deer antler velvet extract is among one of the most disputed supplements in the industry. Many people claim it to have the compound *IGF-1 (Insulin like growth-factor).* While it is noted that the antlers themselves do contain *IGF-1* and some testosterone compounds at their prime of growth, it is difficult to harvest them and extract *IGF-1* while at the same time making it be bio-available for the body to absorb.

However, there are companies out there that have done self-sponsored lab trials; they claim that particular extraction methods have been effective at refining and extracting the active components. Generally, these companies are not cheap; a month's supply can be anywhere from $159-$200. Anecdotal reports have claimed that these products were extremely beneficial in increasing physical performance.
Some of the contents found are in the extract are;

- **Glucosamine Sulfate**, **Chondroitin,** and **Collagen—Aid** connective tissue health such as cartilage and tendons. (Reference those pages for in-depth benefits.)
- It has also been seen to reduce inflammation by reducing *Tumor Necrosis Factor α* and *Interleukin-6 cytokine (inflammatory cell messengers)* expression while promoting *phagocytosis (the process of phagocytes eating invaders).*

In regards to physical performance, deer velvet extract has mixed results with inconclusive data. On the one hand, some small reports suggest performance effects, while others find none; improved recovery from a mineral matrix is not established either. It should also be noted that *IGF-1* has been found in some extracts, most if not all are at trace amounts that would show no significant change to body composition. However, other companies have made the mention that it is the growth factor matrix, not *IGF-1*, that promotes athleticism. This statement refers to the other growth factors found in the Antler Velvet that can also be found in the human body, such as *fibroblast growth factor, nerve growth factor, IGF-II* and *transforming growth factor β.* Again, scientific data is highly lacking and inconclusive; however, there is various anecdotal evidence to support some of these claims.

Proposed benefits (evidence varies)

- May promote joint health
- May improve immune function
- May improve athleticism

Possible Side Effects

For lack of data from research or user reports, no common side effects have been recorded.

Dosage

- 1-6 grams of extract a day, lower for general health, higher for athletic performance

WARNING

Seek medical advice before use. Deer Antler Velvet Extract can affect those on **hormone therapy** such as, **endometriosis, uterine fibroids, breast, ovarian, and uterine cancer**, as it has been seen to contain trace amounts of various hormones.

Desiccated Liver (Defatted liver)

Desiccated Liver does not have much clinical research behind it; generally, this product can be seen sold as desiccated liver tablets or pills. Even though there are many products out there with limited to no clinical data to support extravagant claims, the same caution applies to desiccated liver tablets. Another thing is that they are rather cheap for a month's supply, anything over $20 would be pushing it. Desiccated Liver tablets have a long history in bodybuilding and have been used by various legendary individuals. The main marketing seen for Desiccated Liver is that they are a good nutritional supplement.

Outside of a nutrition supplement, desiccated liver contains a decent amino acid profile, having all nine essential amino acids. The content amount can vary depending on the method used in the desiccation/defatting method.

Nutrition Content	1 g or 1,000 mg	Nutrition Content	1 g or 1,000 mg
Valine	- 9.41 mg	Threonine	- 3.49 mg
Methionine	- 4.35 mg	Isoleucine	- 5.85 mg
Leucine	- 16.75 mg	Phenylalanine	- 9.84 mg
Histidine	- 4.12 mg	Lysine	- 14.43 mg
Arginine	- 13.03 mg		

It has also been found to have a mixture of *Monounsaturated fatty acids (MUFA)* and *Polyunsaturated fatty acids (PUFA); MUFAs* have been seen to reduce *systolic (maximum blood pressure in one heartbeat)* and *diastolic (minimum pressure between two heartbeats)* blood pressure. *PUFAs* are involved in the regulation of inflammation, especially in the regulation of chronic inflammation. Additionally, the antioxidant *Glutathione (an endogenously created antioxidant)* can promote nutritional clearance (after a meal), as it is able to reduce dietary *lipid peroxidation* through the reduction of oxidants found in the food that can destroy nutritional profile. This reduction in lipid peroxidation can also reduce the number of oxidants that enter the body through the digestive tract.

- Iron plays a key role in *hemeproteins (e.g., Hemoglobin, myoglobin)*. Hemoglobin is able to carry oxygen and nutrients more effectively. H*eme-bound* iron is generally absorbed more efficiently than *non-heme iron, but absorption varies; these terms do not mean natural versus synthetic.*

Vitamin B-12 plays a key role in the metabolization of fatty acids and amino acids; it is also involved in the *methylation (adding a methyl group)* synthesis of neurotransmitters

Proposed benefits (evidence varies)

- Enhance dietary benefits
- Promote cardiovascular health
- Improve performance

Possible Side Effects

From a lack of clinical data, there are no known or reported side effects. Read the warning below.

Dosage

- There is no set dosage; 500 mg a day with a meal has shown nutritional benefits & absorption

WARNING

Seek medical advice before use. Desiccated Liver tablets for people with **genetic hemochromatosis** as iron can build up in the body to an overdose level. The nutrient table is illustrative, not a verified specification for every product. Excess Vitamin A and iron matter; diagnosed deficiency needs assessment.

Diindolylmethane *(3, 3'diindolylmethane)*

Diindolylmethane (DIM) is a compound that can be found in *Brassicaceae* vegetables such as broccoli, brussels sprouts, cabbage, and kale. Brussels sprouts contain the highest concentration of DIM. The main reason these dark green vegetables have the reputation of being healthy is because of this very compound. DIM has been seen as a positive modulator of estrogen, keeping it in balance, not too little and not too much.

- Heating or cooking these vegetables reduces the bioavailability of the active compounds reducing its health benefits. Do not confuse it with Indole-3-Carbinol; this is the precursor before it is further metabolized into DIM. High doses have been associated with an increase of *malondialdehyde (a marker of oxidative stress) in the testes, which can reduce testosterone.*

When consumed, the enzyme *myrosinase (plant enzyme)* breaks down *glucobrassicin (derived from glucose and amino acid)* into *Indole-3-carbinol* which gets further converted into DIM. Most of the bioactivity/benefit does come from DIM, while some effect is seen through the precursors; DIM has more molecular targets.

- DIM can target and activate *CYP1A1* and *CYP1A2 (both of these are a type of cytochrome P450).* This can increase *cytochrome P450* enzyme expression aiding the liver in detoxification as *CYP1A1* and *2* are involved in *Phase I* metabolism involving the break-down of drugs and *xenobiotics (unnatural substances).*
- *CYP1A2* can also enhance the synthesis of testosterone by breaking down *estradiol (a stronger form of estrogen)* through *2-hydroxylation (the breakdown of)* into the less powerful/effective *2-hydroxyestradiol* as it can weakly bind to estrogen receptors.

DIM seems to stimulate *ataxia-telangiectasia mutated kinase (ATM); this* is an enzyme that regulates DNA repair and *apoptosis (programmed cell death)* in response to genetic damage by inhibiting *Protein Phosphatase 2 (PP2A)* generally looked at as a tumor suppressor, this enzyme inhibits the activation of ATM, at lower doses. With the additions of the *hydroxylation of estradiol*, this could possibly aid at the reduction of estrogen-induced cancerous and benign tumors such as breast cancer and endometriosis.

- *PP2A* makes up part of *ATMs* structure keeping it inactive until DNA damage is present.

Benefits

- Anti-cancerous
- Promote DNA repair
- Promote liver detox

Possible Side Effects

- Stomach discomfort
- Gas
- Bloating

Dosage

- 100 mg of the standardized extract can maintain estrogen levels balanced

WARNING

Please consult with a medical professional prior to administering Diindolylmethane. There is concern that taking high doses of Diindolylmethane, over 600 mg, can reduce sodium levels in the body. There is concern that Diindolylmethane can affect those on **hormone therapy** or those with **hormone-sensitive conditions** such as **endometriosis**, **uterine fibroids**, **uterine**, **ovarian and breast cancer**, as it can interact with hormone levels.

DMAE *(Dimethylaminoethanol/2-Dimethylaminoethanol)*

DMAE has mixed and inconclusive results when it comes to testing; various anecdotal results seem to benefit its users. DMAE can be found marketed as a nootropic although limited older trials do not establish routine cognitive enhancement, including in older adults. It is important to note that the lack of data and reports will leave this supplement with unknowns to include benefits and side effects if there are any.

Here is the data that is known;

- *Acetylcholine* is an active compound that directly affects memory, stress and motor coordination; it was believed that DMAE was a precursor to *acetylcholine* which could, in theory, increase the concentrations in the body if that were true. However, DMAE cannot be converted into *acetylcholine*. Its main cognitive effects seem to come from preserving *choline (precursor to acetylcholine)* levels by inhibiting the metabolism of *choline* by peripheral tissues. This increase in free choline in the brain can promote *acetylcholine* production due to the stimulation of the *cholinergic receptor*. Generally, this effect is prominently observed in elderly individuals. Choline is also needed to create **lecithin** (ref page for further benefits) and is a major source of methyl groups needed to create **SAMe** (Ref page for further benefits).
- DMAE is able to promote skin rejuvenation when applied topically on the skin; it is believed that DMAE is able to promote thicker **collagen** fibers and increased *dermal (where collagen fibers are found)* thickness. Anecdotal reports state that fine lines, wrinkles and dark marks are lifted after application. This may be due to DMAE's ability to scavenge and break down lipofuscin *(this is yellow-brown fat-based pigment seen on the skin as we age); however, DMAE* is only able to accomplish this when used with *chlorophenoxyacetic acid,* and cannot achieve this by itself.

DMAE is a new supplement with extremely little research and data. Weigh the benefits with the possible unknowns. What works for some might not work for others. It is important to note that there have been some *negative* results. DMAE intake in some animals has been seen to shorten life span. Some trials have also shown the topical application did improve skin appearance, but cells stopped dividing and after 24 hours, a portion of them died.

Proposed benefits (evidence varies)

- Promote cognitive function
- Improve skin appearance
- Enhanced recovery

Possible Side Effects

- Insomnia
- Headaches
- Drowsiness

Dosage

If using an active compound, follow instruction labels diligently;

- Dosage may vary, 130 mg of active DMAE has anecdotal benefits
- Dosage of 200-300 mg has been used as a nootropic supplement, and to aid with ADHD symptoms

WARNING

Seek medical advice before use. DMAE can cause an allergic reaction in those allergic to **Para-Amino-Benzoic-Acid** (PABA) as DMAE has structural similarities to PABA.

<u>Dong Quai *(Angelica sinensis)*</u>

<u>DO NOT CONFUSE THIS HERB WITH DANG QUAI</u> (reference **Dang Quai**) although similar in name, they are two different herbs used by two different groups of people for different health benefits. Confusing one with the other could lead to adverse health effects. Dong Quai is the herb from the *Apiaceae* family, where carrots and celery can be found. Dong Quai has been used in traditional Chinese medicine to improve cardiovascular health and enhance the immune system.

Angelan, the bioactive component in Dong Quai, is thought to enhance immune function by improving *cytokine (cell signaling)* function of white blood cells, thereby stimulating the production of *interferon* and *interleukin-6* macrophages.

- In short, interferons are little messengers in cells that are sent out after a virus (bacteria, tumor, etc.) attacks a cell; these messengers tell other cells around to heighten their defenses. Interleukin-6 is a messenger of *macrophages* and *T lymphocytes* (types of white blood cells) that is responsible for inflammation/defense response to the area that was attacked (hurt, burnt, damaged, etc.) thereby increasing immune response and blood flow.

This mechanism is mainly achieved through the ability to improve circulation through the enhancement of *Inducible Nitric Oxide Synthase (iNOS). iNOS* is an enzyme that is involved in producing *Nitric Oxide (responsible for vasodilation/blood flow)* for innate immune response, especially in pro-inflammatory signals. *Acidic Angelica Polysaccharide (AAP)* has been seen to activate *Toll Like Receptor 4 (gene produced protein that leads to innate immune responses)* in *macrophages (type of white blood cells)* to induce *iNOS* and *Tumor Necrosis Factor Alpha (TNF-α is a cytokine/messenger involved in inflammation)*, basically meaning *AAP* is able to cause *macrophages* to create *iNOS* via the signaling of TNF-α; this can promote blood flow to an area aiding immune responses like *phagocytosis (process of phagocytes killing invaders)*.

- *Ferulic acid*, a polyphenol found in Dong Quai, has been seen to reduce *thromboxane* B_2 production. *Thromboxane 2* can facilitate platelet aggregation as well as be a vasoconstrictor *(constriction of blood vessels and vascular muscle cells)*. This can promote *Endothelial NOS*, potentially reducing clots.

<u>Proposed benefits (evidence varies)</u>

- Immune system boost
- Vasodilator
- Antioxidant
- Cardiovascular health aid

<u>Possible Side Effects</u>

- Photosensitivity
- Low Blood Pressure
- Mild Sedation

<u>Dosage</u>

No set standard dosage is known; adhere to label directions. The following is based on research:

- For general health and vasodilation 1-1.5 grams 3x a day

<u>***WARNING***</u>

Seek medical advice before use. Review **autoimmune disorders,** including **multiple sclerosis**, **lupus**, **rheumatoid arthritis** and **pemphigus vulgaris,** because of proposed immune effects. Avoid in pregnancy; review bleeding disorders, *<u>warfarin and other anticoagulants. Menstrual benefits are unestablished. Angelica sinensis and Angelica gigas are different herbs, with non-interchangeable effects and doses.</u>*

Ecdysterone *(Ecdysteroid)*

Ecdysterone is a type of *ecdysteroid* along with **turkesterone**. *Ecdysteroids* are insect *steroid hormones* that are involved in molting and in the reproduction of insects. They are also found in some plants that create *ecdysteroids* in response to environmental stress. Currently they are marketed to enhance muscle mass while also increasing general health in humans. *Ecdysteroids* are structurally similar to *androgens* but structural similarity does not establish steroid-equivalent effects, absence of hormonal effects or *long*-term safety in humans.

The "claimed" side effects are the benefits it's used for. *Ecdysteroids' anabolic* activity comes from enhanced protein synthesis in muscle tissue. Preclinical comparisons with anabolic agents do not establish comparable muscle-building effects in pe**ople**.

- There are various mechanisms in which it is able to accomplish this. One being through enhanced *mRNA Translation* directly geared towards enhanced protein synthesis, since protein synthesis is regulated during transcription and translation of *RNA*. This has been seen to happen in the *ribosomes (location in the cell where protein synthesis happens)* of *eukaryotic (cells with a nucleus and membranes)* cells.

Protein synthesis comes from *RNA Translation,* which uses an *mRNA* sequence as a template (kind of like a recipe/instructions) to create a specific amino acid chain (the product it wants to build) based on its genetic code (rules for protein synthesis to create the specific product), in this case, it produces/enhances specific peptides/polypeptides used in protein synthesis needed for muscle repair.

- The second pathway has been seen to cause a Calcium ion influx into *myocytes (muscle cells)* which can cause *phosphorylation* of *AKT* which ultimately signals protein synthesis. The *AKT pathway* promotes cell growth and responds to *extracellular (outside of the cell)* signals, such as a Calcium ion influx.
- The third pathway is believed that ecdysteroids can increase the size of myotubes (muscle fibers) through interacting with the *α* and *β Estrogen Receptor* as activation of these receptors has been linked to an increase of *MyoD (a protein involved in muscle tissue formation).*

Additionally, *Ecdysteroids* are able to support *neurons* by protecting them against *neurological* toxins as it can promote the secretion of *acetylcholinesterase* and *glutamic decarboxylase.* These are enzymes in charge of breaking down overproduced neurotransmitters that can cause toxicity to neurons.

Proposed benefits (evidence varies)

- Promote lean muscle
- Does not affect hormones
- Can reduce blood sugar
- Enhance endurance
- Aid fat loss
- Antioxidant

Possible Side Effects

There currently have been no reported side effects

Dosage

- Human evidence is limited and mixed; no standard effective and safe bodybuilding dose is established.

WARNING

Seek medical advice before use. Although there have been no known negative side effects reported, research is still lacking. If you have a **metabolic disorder, seek** medical advice before taking any ecdysteroid.

Echinacea

Echinacea is a group of flowering plants from the *Asteraceae* family where daisies and sunflowers can be found; they are native to North America and can be predominately spotted in open woods or prairies. Echinacea has been used in traditional folk medicine in the prevention and alleviation of colds. There are various echinacea species, *Echinacea purpurea* and *Echinacea angustifolia* being predominantly used for their high bioactive content. Any reduction in catching colds may be modest; evidence for shortening colds is unclear and preparation-specific.

The main benefits seen in Echinacea extracts and supplements seem to come from *alkamides, polysaccharides and caffeic acids (caffeic acids seem to have poor bioavailability and may not contribute to immune benefits)* that have immunostimulant activity. It is able to promote *leukocyte (white blood cells)* activity, through activating the white blood cells: *neutrophils, macrophages* and *natural killer* cells.

- *Alkamides* have been seen to bind to *Cannabinoid Receptors 1 & 2; these* receptors have been seen to increase *Nuclear Factor-kappa Beta (NF-kB is involved in cell death and survival and DNA transcription).* This has been seen to positively modulate immune responses only in the presence of *pathogens (foreign harmful substances); this* may be a reason for its *leukocyte* benefit.
- The *polysaccharides* in the plant have been shown to increase *Interleukin-1* & *6 (IL-1 & IL-6)* and *Tumor Necrosis Factor-α* expression by *macrophages.* These inflammatory *cytokines (messengers)* regulate inflammatory and immune responses resulting from infections or damage to the body.

These actions can augment *phagocytosis,* the process in which a *phagocyte* eats foreign contaminants, e.g. Bacteria, viruses, etc. as well as reduce excessive inflammation brought on by infections or insults to the body. Recent research also shows that the *Alkamides* within Echinacea may have *anxiolytic* properties since it has a similar structure to *anandamide,* which affects anxiety. Through this, it is able to aid reduction in anxiety. More research is needed to confirm the benefit and mechanisms.

Proposed benefits (evidence varies)

- May reduce anxiety at lower dosage
- Enhance/support immune function

Possible Side Effects

- Abdominal pain
- Erythema
- Nausea
- Rash

Dosage

- The recommended dosage of a standardized extract of Echinacea purpurea is 900-1,500 mg a day and can be broken into 3 doses throughout the day.

WARNING

Seek medical advice before use. Echinacea can affect people with **auto-immune disorders** such as; **multiple sclerosis**, **lupus**, **rheumatoid arthritis** or **pemphigus vulgaris** as Echinacea has been known to affect the immune system which may worsen the condition. If you have **ragweed** allergies, Echinacea supplements could cause an allergic reaction since Ragweed is from the *Asteraceae* family like Echinacea.

Elderberry *(Sambucus nigra)*

Elderberry, also called elderflower, has been used to treat wounds, through topical application, and illnesses like the cold and flu. While it may not prevent or heal colds or the flu, small studies suggest possible symptom effects, but reliable clinical evidence is limited and infection treatment is not established. It seems to aid respiratory illnesses by reducing swelling in mucous membranes which could aid in sinus symptoms. **DO NOT EAT** raw elderberries as they can be **POISONOUS**. Elderberries contain *cyanogenic glucosides*, a chemical similar to cyanide.

The combination of elderberry and echinacea extract has synergistic components, enhancing each other's effects, possibly potentiating their antiviral properties. The active components of elderberry are the matrix of *flavonols; quercetin-3-glucoside, quercetin-3-rutinoside,* and *anthocyanin.* Laboratory elderberry extracts have shown activity against some *Gram-positive* and *Gram-negative bacteria; this does not establish treatment in humans.* The extract also seems to be able to inhibit influenza virus multiplication. It seems to accomplish this through:

- *Flavonols* as they seem to have the ability to block bacterial adhesion, making it difficult for bacteria to stay and infect. It has been observed that the *flavanols* content of elderberry extracts can inhibit the *neuraminidase* effect of influenza on healthy cells.
    - The *neuraminidase* effect, referred to as *viral neuraminidase* for viruses, is the mechanism that is needed for virus/bacterial replication. *Neuraminidases are enzymes* that are used to "cut through" molecules that line the host cell; this is needed to attach and exit the host cell. In simpler terms, they prevent the virus from attaching themselves to a cell in the body and starting to multiply.

Anthocyanins, found in Elderberries, have also been shown to have antioxidant properties as it has been shown to reduce excessive levels of *Interleukin-6 (a pro-inflammatory cytokine)* as well as maintaining proper *Interleukin-15 (IL-15). IL-15* has been implicated in obesity-related inflammation and diminished glucose function as it regulates activation and *proliferation (cell multiplication)* of *natural killer (NK)* cells, specifically *CD56*^{bright} *NK* cells which have been seen to mainly produce proinflammatory *cytokines*.

- The anti-hypertension/vasodilator effects are currently not known, although it is believed that this effect is mediated through its anti-inflammatory/antioxidant effect.

Proposed benefits (evidence varies)

- Enhance immune system
- Reduce cold symptoms
- Anti-viral
- Reduce flu symptoms

Possible Side Effects (Top row is from raw Elderberry consumption)

- Nausea
- Vomiting
- Diarrhea
- Low Blood Pressure

Dosage

Dosage varies by extract concentration. Adhere to labels instructions.

WARNING

Seek medical advice before use. Elderberries can worsen **autoimmune diseases** like **multiple sclerosis**, **lupus**, **rheumatoid arthritis** or **pemphigus vulgaris** as they can affect immune function. Proper processing matters: raw/unripe berries, leaves and stems can contain cyanogenic compounds; do not assume all home preparations are safe.

Evening Primrose Oil *(Oenothera biennis)*

Evening Primrose, also known as German Rampion or the Fever Plant, is a flowering plant from the *Onagraceae* family with bright yellow flowers that is native to North America and Europe. Generally, *Oenothera* plants (or genus) have been seen to have a lot of beneficial bio-activity, *Oenothera biennis* being the most commonly known. Evening Primrose seed oil has been shown to contain the highest concentration of bioactive compounds compared to the rest of the plant as the oil contains the omega-6 fatty acids, *linoleic acid* and **Gamma-linolenic acid (GLA), which are** precursors to the anti-inflammatory *eicosanoids (signaling molecules). Linoleic acid is essential; the body can convert it into GLA and other omega-6 fatty acids.* Generally, it can be seen marketed as a supplement to aid skin aesthetics and to obtain omega fatty acids. The entire plant is edible, from the roots all the way to the flower petals. Current evidence does not establish clinical benefit for the conditions commonly promoted, despite the fatty-acid mechanisms.

The oil is mainly used to aid with illnesses or problems with inflammation; other uses such as breast pain and PMS symptoms have not had evidence shown to prove its efficacy. The main benefit of this plant does come from its fatty acid content, mainly GLA. In the body, GLA gets converted into *dihomo-gamma-linolenic acid (DGLA)* by *elongase-5*. Here *DGLA* can further be oxidized into *15-hydroxyeicosatrienoic acid (15-HETre)* by *lipoxygenase (an enzyme that converts fatty acids)* or metabolized into *Prostaglandin E1* by *cyclooxygenase (involved in the creation of inflammatory mediators)*, both of which have anti-inflammatory properties.

- *15-HETrE* can block the oxidation of **arachidonic acid (AA)** into *Leukotriene A4 (LTA4)* by inhibiting *5-lipoxygenase (5-LOX); 5-LOX* converts AA into *leukotrienes. Leukotrienes* are involved in inflammatory and allergic conditions as well as *vasoconstriction (constriction of blood vessels), LTA4* has been seen as the basis for the creation of pro-inflammatory *leukotrienes.*
- *Prostaglandin E1* has been seen as a vasodilator; it has also been seen to have *anti-thrombotic (anti-clotting)* effects by activating *Prostaglandin receptor; this* can lead to increased *cyclic Adenosine Monophosphate* levels in smooth muscles resulting in lower Ca^{2+} *(Calcium Ion)* levels within the cells allowing relaxation of blood vessels, since *Ca2+ is* involved in muscle contraction.

Proposed benefits (evidence varies)

- Cardiovascular health aid
- Vasodilator
- Anti-inflammatory
- Promote skin health (through its anti-inflammatory effects)

Possible Side Effects

- Low blood pressure
- Diarrhea
- Headache
- Upset stomach
- Headache
- Nausea

Dosage

- The recommended dosage is 1-3 grams a day

WARNING

Seek medical advice before use. Evening Primrose Oil may affect those with **Bleeding Disorders** as Primrose Oil can slow blood clotting. There is concern that Primrose Oil can increase the frequency of **seizures**.

Evodiamine *(Evodia rutaecarpa/Tetradium ruticarpum)*

Evodiamine is an extract derived from the Evodia Tree from the *Rutaceae* family where citruses can be found. The fruits have a bitter taste and have been traced back to Traditional Chinese Medicine to warm the body in cold weather. It can mainly be seen marketed as a Thermogenic to aid fat loss, since it seems to act in a similar fashion to **Capsaicin**. The isolated compounds have been seen to have poor bio-availability so an extract is recommended or the consumption of the whole berry if possible.

Evodiamine, like Capsaicin just in a weaker fashion, acts via adrenaline receptors *Transient Receptor Potential Cation Channels Subfamily Vanilloid (TRPV).* These receptors are known to be modulated/activated via *vanilloid compounds; however, dehydroevodiamine (a bioactive component of Evodia)* and evodiamine are *quinazoline alkaloids* which are chemically structured differently; this could explain its weaker activation of *TRPV1*. The current activation mechanism is not yet fully understood.

- *TRPV1* is involved in the regulation of body heat and the sensation of heat as it is sensitive and responsive to heat that is experienced. Evodiamine can lower the *TRPV1* threshold of heat needed to be activated all the way down to current room temperature. **Arachidonic Acid (AA)** is also a known agonist; it is believed that one can experience this effect through symptoms from a cold as AA is released from white blood cells.

PGC-1α has been observed to be increased through *TRPV1* activation. *PGC-1α* is the master regulator of *mitochondrial biogenesis; this* increases mitochondrial mass in a cell, which can increase physical endurance. *PGC-1α* has also been seen to promote the secretion of *β-aminoisobutyric acid* which has been linked to thermogenic gene activation in white fat, resulting in additional fat loss. *TRPV1* activation causes a calcium influx; calcium influxes are strong cellular signalers. Cellular calcium influxes may be the reason for noticed athletic improvements observed. Enhanced endurance may also be increased due to the use of fatty acids as fuel resulting from *adrenergic* activity in the release of *catecholamines (adrenaline, noradrenaline, dopamine)* down cascade of *TRPV1* activation.

- *TRPV1 stimulation can increase pain sensation in neuropathic pain since TRPV1 modulates body heat and pain. It is recommended to supplement creatine as well to reduce the possibility of calcium-induced cytotoxicity (cell death) due to the calcium influx caused from TRPV1 activation.* Anti-inflammatory and *analgesic* effects seem to come from *Capsaicin* inhibiting *IkB (Reference* ***Boswellia serrata*** *on IkB for a deeper explanation).*

Benefits

- Thermogenic
- Aid fat loss
- Promote endurance

Possible Side Effects

For lack of data or anecdotal reports, possible side effects have not been established at this time.

Dosage

There currently is no standard dosage; traditional use is the boiling and drinking of 3 to 9 grams of berries 2x a day. Adhere to label directions.

WARNING

Please consult with a medical professional prior to administering Evodiamine. There is concern that Evodiamine can slow blood clotting affecting those with **bleeding disorders.** Evodiamine may affect **pregnancies** as Evodiamine has been seen to affect uterus contractions and litter sizes in animals.

Fenugreek (*Trigonella foenum-graecum*)

Fenugreek is a plant from the *Fabaceae* family, more commonly known as the bean family and can be found growing around the world; it has been traditionally used in traditional Chinese and Ayurvedic Medicine to raise libido. It is mainly seen marketed as an antidiabetic and a testosterone booster. Although there currently are conflicting results in fenugreek's testosterone-boosting benefits, it should be noted that some studies have shown increases in free testosterone levels.

4-hydroxyisoleucine (4-OH I1e), exclusive to fenugreek, has shown protection on *beta cells (store and release insulin)* as it has been seen to induce insulin secretion, especially in high glucose environments thereby reducing circulating glucose levels. It is believed that this mechanism is accomplished by *4-OH I1e* activating *Insulin Receptor Substrate 1* and augmenting *glucose*-induced insulin release, which may enhance insulin sensitivity. *Galactomannan,* found in fenugreek, has been seen to inhibit *alpha-amylase* and maltase enzymes; these enzymes are involved in the breakdown of carbohydrates into glucose for absorption.

- Not only is this beneficial for **diabetics**, but it can enhance endurance training for athletes as well as provide a food partition effect after intense training bouts.

*Diosgenin (similar structure to **DHEA**)* is another active component found in fenugreek seeds. It has been seen to have anabolic activities in muscle tissue. This might be the result of their ability to stimulate the free flow of testosterone in the body; it is believed to accomplish this through *5α-reductase (converts testosterone into dihydrotestosterone)* and *aromatase (converts testosterone into estradiol)* inhibition. Both of these hormones can have negative effects on the male body when testosterone ratios become unbalanced. *Galactomannan* has also been implicated in enhanced muscle growth, although its mechanism is not fully known although it is believed to promote a favorable *Blood Urea Nitrogen (environment involved in muscle building and breakdown)* environment.

- The general increase of testosterone in the body can lead to enhanced results in training. Along with the fact that it can enhance insulin activity, it can aid in higher performance results by enhancing **ATP**, allowing for more effective training sessions.

Benefits

- Increased lean muscle
- Increase overall strength
- Fat loss aid
- Increase free testosterone
- May improve power/explosion

Possible Side Effects

- Gastrointestinal discomfort
- Stomach aches
- Bloating
- "Maple syrup" urine odor
- Congestion

Dosage

- Study recommended dosage is 500-600 mg of fenugreek extract a day.

WARNING

Please consult with a medical professional prior to administering Fenugreek extract. There is concern that Fenugreek can affect those diagnosed with **Diabetes** as fenugreek can affect blood sugar levels; medication may need to be modified. If you have a **Fabaceae Allergy**, you might be allergic to Fenugreek as it is part of the Fabaceae family.

Fish Oil

Fish oil is high in *polyunsaturated omega-3 fatty acids*, containing *Docosahexaenoic Acid (DHA)* and *Eicosapentaenoic Acid (EPA); it* can be commonly found marketed as a joint health aid, *nootropic (cognitive aid)* and to reduce the risk of *Cardiovascular Disease.* Ordinary fish-oil supplements do not reliably prevent *cardiovascular* events; selected prescription products have different evidence. The body converts some essential ALA to EPA and DHA, but conversion is limited; their effects are not interchangeable. *DHA* has nootropic effects that are further enhanced by *EPA's* ability to increase blood flow to the brain; this may very well aid those diagnosed with cognitive problems. It is important to note, those who regularly DO NOT eat fish as part of their diet may see an increase in cognitive functions; for those that DO, benefits *may not be seen from this supplement*.

EPA's and DHA's anti-inflammatory effects may be due to the increase of *EPA* concentrations in *Neutrophils* following supplementation. *EPA* seems to *attenuate (slow/reduce) leukotriene B4 (inflammatory signaler)* production in *neutrophils (a type of white blood cell).* It is believed that it accomplishes this through the suppression of gene transcription for *phosphoinositide 3-kinase (PI3K)* activation. *PIK3* controls intra-cellular signaling and modulation of lipids like **arachidonic acid (AA)**. Modulating the production of *eicosanoids* can either inhibit or cause inflammation.

- *Leukotrienes* are a family of *eicosanoids; eicosanoids* are signaling molecules released from AA oxidation that are involved in inflammatory responses.

This anti-inflammatory reduction is also seen in the brain and may be the reason for its cognitive benefits as pro-inflammatory *cytokines (messengers)*, like *interleukin-1β*, have been seen in neural tissue of stressed and depressed individuals. Inflammation in neurons can decrease *neurogenesis (creation of nerves).* A secondary mechanism is believed through G_s *alpha subunit (a G Protein that activates and is involved in neurotransmission)* modifying, allowing *G proteins* to more effectively activate receptors on the cell. This *G Protein* has been seen to be an incomplete complex in depressed individuals.

Fish oil has been seen to have the ability to enhance protein synthesis, although the mechanism isn't fully known; it is believed to inhibit *myostatin* (*GDF-8*) as Fish Oil has been seen to induce *Mothers Against Decapentaplegic Homolog 7 (SMAD7)* which is believed to inhibit *GDF-8*.

- *Myostatin*, in short, is a growth factor; it is a myokine (a protein) that is released by *myocytes* that directly signal the inhibition of *myogenesis* (creation of muscle tissue).

Proposed benefits (evidence varies)

- May promote muscle growth
- Improve cognitive function
- Anti-inflammatory
- Promotes joint health
- Possible cardiovascular health aid

Possible Side Effects

- May reduce clotting abilities
- Low blood pressure
- Fish "burps"
- Nausea

Dosage

- Product amounts are not the same as EPA+DHA amounts. High-dose treatment of triglycerides uses prescription regimens; routine six-dose use is not established.

WARNING

Seek medical advice before use. Fish Oil can affect those with **low blood pressure** as Fish Oil has been seen to further lower blood pressure. High doses have increased atrial-fibrillation risk in some trials; discuss use with relevant medicines or heart conditions.

<u>Flaxseed Oil/Linseed Oil *(Linum usitatissimum)*</u>

Flaxseed oil is a golden yellowish oil that is derived from the Flax plant in the *Linaceae* family; these plants are native to tropical areas throughout the world. Flaxseed Oil has been traditionally used for drying and varnishing wood in Roman times. As a supplement, it is considered a safe and edible oil, containing the *Polyunsaturated* Omega-3 *fatty acid alpha-linolenic acid (ALA).* Flaxseed Oil also contains various *omega-6 fatty acids* which can hinder *ALA* conversion into its usable form of *omega-3 fatty acids, Docosahexaenoic Acid (DHA)* and *Eicosapentaenoic Acid (EPA).* Supplementation may aid in supporting the cardiovascular and immune systems as well as provide brain health support.

ALA can be converted into *eicosapentaenoic acid (EPA)* and then *docosahexaenoic acid (DHA).* The process, however, is extremely limited as *omega-6 fatty acids* compete for the same enzymes needed to convert *ALA* to *EPA* and *DHA.* Generally, *ALA* is converted by the enzyme *Delta-6 desaturase (needed for omega-6 conversion), elongated* by *ELOVL fatty acid elongase 5,* then converted once more by *Delta-5 desaturase (needed for omega-6 conversion)* to successfully create *EPA.* To further create *DHA,* one of two processes can occur, one longer one shorter. The shorter one, *EPA* is *elongated* by *ELOVL fatty acid elongase 5* and then converted by *Delta-4 desaturase (needed for omega-6 conversion)* into *DHA.* The process is a rather complex one, and this is the reason for the lack of **omega-3** acids in Vegans and Vegetarians compared to those who consume Fish Oil. ALA-rich oil is not clinically interchangeable with EPA/DHA products; larger ALA doses do not guarantee equivalent tissue levels or outcomes. For full health benefits of *DHA* and *EPA* refer to **Fish Oil** or Omega-3.

Whole flaxseed contains lignans that have been seen in preclinical research to reduce *Estrogen Receptor β* expression and inhibit *aromatization* and *17β-hydroxysteroid dehydrogenase* which are involved in the conversion of estrogen due to *enterolactone,* a *phytoestrogen lignan* that is perceived by the body as a weaker version of estrogen. *Lignan secoisolariciresinol diglucoside (SDG)* may also contribute to the reduction of tumor growth by decreasing the production of tumor cells and increasing *apoptosis (programmed cell death)* through modifying estrogen and growth factor signals. *Lignan (SDG),* has been shown to inhibit glucose synthesis in the liver which may lead to reduced glucose levels. Ordinary refined flaxseed oil contains little lignan or fiber unless these are added; whole-seed findings cannot be assumed for the oil.

<u>Benefits</u>

- Possible anti-carcinogen
- Vasodilator
- Anti-inflammatory
- Cardiovascular health aid
- Stabilize blood sugar levels

<u>Possible Side Effects</u>

- Diarrhea
- Low blood pressure
- Gastrointestinal discomfort

<u>Dosage</u>

- For general health 7 grams a day of flaxseed oil
- For diabetes 13 grams of flaxseed oil a day or 32 grams of milled flaxseed

<u>***WARNING***</u>

Seek medical advice before use. Flaxseed Oil can affect people with a **bleeding disorder**, as Flaxseed Oil can reduce blood clotting. Flaxseed Oil may affect those in **hormone therapy** or t hat have **hormone sensitive** conditions, such as **breast cancer** or **ovarian cancer** due to its phytoestrogen content.

Folic Acid (Vitamin B9)

Folic acid is the synthetic form of Folate, and either the natural or synthetic form is an extremely important and essential vitamin for general health and pregnant women. This is vital for pregnant women as it helps prevent *neural tube defects; these* are birth defects that often develop within the first month of pregnancy; the most common ones are *spina bifida (spinal cords do not develop properly)* and *anencephaly (born with an underdeveloped brain and possibly an incomplete skull).* *These birth defects generally happen within the first 28 days of conception. It is vital to have an adequate intake of Folic Acid as these birth defects can affect life outcomes and can lead to paralysis of the lower body.* Pregnancy folate needs remain higher throughout pregnancy; do not stop prescribed prenatal folic acid after the first trimester on the basis of uncertain allergy associations.

Folic acid is good for general health, as it is involved in *methylation* and *DNA Synthesis. Methylation* plays a huge role in various processes in the body from the creation of neurotransmitters to RNA processing and DNA replication; both are key factors in embryonic development. The body converts the inactive Folate into *L-methylfolate* by the enzymes *dihydrofolate reductase* and *methylenetetrahydrofolate reductase*. These conversions provide *methyl* donations needed for RNA and DNA creation/processes.
Faulty *methylation* can cause various abnormal conditions in the body like Chronic Fatigue Syndrome and anemia in mothers, and like mentioned before, cognitive and physical problems in fetuses.

- While there seem to be no side effects from moderate folic acid consumption, there is concern that it can mask a **Vitamin B-12** deficiency. Vitamin B-12, Folate and Iron can effectively treat anemia due to blood volume expansion as a result of pregnancy as they are required for *erythropoiesis (creation of red blood cells in the bone marrow.)*

Folic acid intake can also reduce the risk of depression after birth by aiding ***SAMe (S-Adenosyl methionine)*** in *transmethylation.* SAMe's antidepressant effects seem to be derived from methyl group donation to *catecholamines* that affect mood. *Catecholamines* are a family of chemicals to include *dopamine, adrenaline,* and *noradrenaline.* An enhanced methylation process can promote *dopaminergic* activity while also increasing the *serotonin* rate of replacement leading to an "uplifted" mood. These chemicals are involved in responses, such as happiness, sex, joy, hunger, etc.

Benefits

- Aid in healthier pregnancy
- General good health

Possible Side Effects

- May mask B-12 deficiency
- Upset stomach
- Nausea
- Abdominal cramps
- Possible mood changes

Dosage

- Adult RDA: 400 mcg dietary folate equivalents (DFE)/day; pregnancy 600 mcg DFE. Those who could become pregnant should obtain 400 mcg folic acid daily in addition to food folate.

WARNING

Seek medical advice before **use**. Folic Acid can mask **B-12 deficiencies; this** can lead to neurological damage to fetuses and anemia to individuals.

Forskohlii (*Plectranthus barbatus*)

Coleus Forskohlii, also referred to as *Forskolin* the bioactive compound, is from the *Lamiaceae* family where mint and lavender can be found. It's been known to grow in tropical climates and has purple-ish bell looking flowers around the stem. It has been traditionally used in Ayurvedic medicine for intestinal, cardiovascular, and nervous system health. It can be found marketed as a fat loss aid and a testosterone booster. <u>Caffeine can inhibit the effects of *forskolin* as they both use the *adenosine receptors.*</u>

Forskolin has been seen as a fat-burning agent with free testosterone boosting properties, which theoretically could improve body composition. It is able to accomplish this as *Forskolin* has been seen to activate all forms of *Adenylyl Cyclase (this range of activation could lead to side effects), which aids* in the conversion of **ATP** into *cAMP (Cyclic Adenosine Monophosphate). cAMP* is involved in *intracellular signal transduction (chemical signaling inside the cell)* in practically all cells within the body; this increase in *cAMP* levels is able to increase enzyme activity in a similar fashion to exercise.

- It is believed that this effect is also mirrored in muscle tissue, which could stimulate protein synthesis by activating *Protein Kinase B (involved in cell multiplication/survival/metabolism)* through *phosphorylation (a type of on/off mechanism by adding phosphoryl group)* of the enzyme *Phosphatidylinositol 3-kinases (involved in cell growth/multiplication)* in a similar fashion to exercise.
- It is believed that *cAMP* may also increase testosterone in the same manner as listed above, since *Luteinizing Hormone (LH)* increases *cAMP* levels in *Leydig cells (cells that create testosterone located in the testes).* The increase of *cAMP* in the testicles can mimic the effects of the *Luteinizing Hormone (LH)*; without actually having *LH* be present, through this the testes may increase the free flow of testosterone.
- *Forskolin's fat loss* effects seem to come from *cAMP* modulating *insulin's effects* on muscle mainly by inhibiting *Glucose Transporter Type 4 (insulin-regulated transporter that can be found on muscle/fat tissue)* via the *Mammalian Target of Rapamycin (involved in the activation of insulin receptors)* while activating *Protein Kinase B* in muscle tissue, like mentioned before, which could positively affect *lipid metabolism (as in fat as fuel instead of glucose).*

<u>Benefits</u>

- Increase free testosterone
- Fat loss aid
- Improve body composition

<u>Possible Side Effects</u>

- Low blood pressure
- Headache
- Nausea
- Fatigue
- Dizziness

<u>Dosage</u>

- 250 mg 2x a day of a 10% forskolin extract before a meal for weight loss/testosterone boosting

<u>***WARNING***</u>

Please consult with a medical professional prior to administering Forskolin. There is concern that forskolin can affect those diagnosed with **stomach ulcers** or any stomach disease as it has been seen to affect stomach acid levels. Prolonged concentrations of *cAMP* can affect the synthesis of proteins in an attempt to normalize cAMP levels; this can affect recovery time for individuals, i.e. surgery, skin lesion, muscle, etc. Forskolin can affect **Diabetics** as it can affect insulin levels.

Fucoxanthin *(Undaria pinnatifida/Wakame)*

Fucoxanthin is a xanthophyll *(yellow pigment)* with a similar structure to *carotenes* like Vitamin A, without actual vitamin-like qualities. It is generally derived from brown algae but can be acquired from other seaweed like *Undaria pinnatifida (Wakame)* and like mentioned before, its main purpose is to give the algae its green brownish color. This supplement is mainly geared towards obese people as a therapeutic approach at weight-loss. Generally seen, weight-loss effects from Fucoxanthin are not seen for 5-16 weeks as it has a build-up effect in the body. For those that already have an active or otherwise healthy lifestyle, there doesn't seem to be much weight loss benefit.

Fucoxanthin becomes converted into *fucoxanthinol* in the body, becoming the main active component. *Fucoxanthinol* seems to be highly stored in fat cells, over any other cell in the body. Fucoxanthin has additionally been seen to be implicated in increased *RNA expression* of *adiponectin (a hormone involved in fatty acid breakdown and glucose regulation).*

- This expression can then activate *5' Adenosine Monophosphate-Activated Protein Kinase (AMPK)* in *adipocytes (fat cells)* via *phosphorylation (a type of on/off mechanism by adding phosphoryl group). AMPK* activates fatty acids or glucose usage when cellular energy is low to maintain optimal cellular energy levels. The antidiabetic effects are due to *glucose transporter 4 (GLUT4) RNA expression* in muscle tissue which can result in excess glucose use for energy.

Fucoxanthin has also been seen to reduce expression of *Peroxisome Proliferator-activated Receptor Gamma (PPARγ); PPARγ* is mainly located in fat cells and regulates fatty acid storage; PPARγ expression increases *adipogenesis (fat cell creation)* of fat cells.

- Fucoxanthin seems to be able to stimulate *Uncoupling Protein 1 (UCP1) RNA expression,* in *white adipose (stored fat/white fat can be visually seen). UCP1* is generally seen in *brown adipose (brown fat located around the organs)* and used for heat production to maintain organ heat. However, fucoxanthin is believed to accomplish the same in *white adipose,* thereby stimulating a type of thermogenesis and heating the body up, thereby burning fat during resting energy expenditure.

It can also reduce inflammation by reducing *Nitric Oxide* and inflammatory *cytokines (messengers)* like *Interleukin 1β* and *6* in the body. It is important to note that this can increase blood pressure.

Benefits

- Improved body composition
- Aid in insulin sensitivity
- Enhance fat loss

Possible Side Effects

- High blood pressure

Dosage

- 2.4-8 mg of fucoxanthin a day

WARNING

Please consult with a medical professional prior to administering Fucoxanthin. There is concern that Fucoxanthin may affect those with **blood pressure problems** as Fucoxanthin can reduce nitric oxide, which could increase blood pressure. Fucoxanthin may affect diabetics as it has been seen to affect insulin levels; medication may need to be modified. Not much research has been done behind fucoxanthin, so side effects are not fully known. Early research shows promise to aid with obesity

GABA (Gamma-aminobutyric Acid)

GABA is one of the main *depressive* neurotransmitters in the brain that reduces neuron excitability throughout the nervous system by balancing the neurotransmitter **glutamate.** The fact that GABA is seen as a neuro-downer is not meant in a negative context. It's responsible for depressive actions, such as sedation and relaxation. However, it is important to note that oral GABA supplementation does not reliably establish increased brain GABA; human blood-brain-barrier penetration remains uncertain. Supplementation has been seen to improve sleep quality in some individuals.

GABA is highly regulated within the body; direct external consumption might not present benefits; however, through consumption at peak times may enhance availability (e.g. sleep/exercise). Blood GABA levels are not a reliable direct measure of brain GABA concentrations. GABA can self-inhibit, meaning it can block high doses of GABA into the brain in an attempt to maintain stable levels. It is believed that GABA supplementation may activate *$GABA_A$ receptor* which is associated with *anxiolytic (anti-stress)* and sleep effects. *$GABA_B$* receptor has been associated with wakening and dreaming, while *$GABA_C$* receptor has been associated with deep sleep.

- If somehow there's an excess GABA uptake, the brain has the ability to activate an *efflux (removal)* of GABA that's 16x more potent than the uptake. This means that at smaller regular doses, supplementation may affect GABA brain concentrations.

Through an increase in *nitric oxide*, brain permeability for GABA may be enhanced. Since *nitric oxide* increases during resistance training, GABA might be able to go through the blood-brain barrier. This could be the reason as to why it is used for muscle tone.

- GABA consumption might be able to stimulate *Growth Hormone* secretion, especially if taken during exercise and before sleep.

It's important to note that since GABA is a neurotransmitter, there is not much data or clinical research that can be done currently to conclusively say these things 100%. Human brain function can be studied with imaging and other methods, but oral GABA trials remain small and evidence for stress or sleep benefit is limited. Much of this mechanism is through theory based on the enhancements seen through clinical trials and anecdotal reports.

Proposed benefits (evidence varies)

- May enhance muscle growth
- Sleep aid
- Anxiolytic
- May improve mood
- Nootropic

Possible Side Effects

- Drowsiness
- Shortness of breath
- Low blood pressure
- Paresthesia

Dosage

- No established muscle-building dose exists. Small sleep/stress studies use varied preparations and doses; long-term safety remains uncertain.

WARNING

Seek medical advice before use. Its neurotransmitter role does not establish benefit or safety of an oral supplement.

GLA (Gamma-linolenic Acid)

GLA is a *Polyunsaturated omega-6 fatty acid* that is found in oils such as borage, evening primrose and blackcurrant seed oil. It can also be found in some fish oil supplements as well as a stand-alone supplement. GLA is made from essential linoleic acid; GLA itself is not classed as an essential *fatty acid*. GLA plays a vital role in the synthesis of the anti-inflammatory precursors, *eicosanoids (signaling molecules).* No universal optimal 4:1 ***omega-6:omega-3* ratio is established; overall intake and the fats being replaced matter**. The typical American diet, however, is high in *omega-6 fatty acids*, increasingly so, as the average American diet includes **a 10:1 to a 20:1 *omega-6* to *omega-3 fatty acid* ratio**, quite frequently higher ratios have been reported. If this is the case for your given diet, *omega-3 fatty acids* will benefit your current health. Fish Oil**, Omega-3s** or **flaxseeds** may show promise. Generally, it can be found marketed as a supplement to aid skin aesthetics and as an anti-inflammatory aid.

In the body, GLA gets converted into *dihomo-gamma-linolenic acid (DGLA)* by *elongase-5.* Here *DGLA* can further be oxidized into *15-hydroxyeicosatrienoic acid (15-HETre)* by *lipoxygenase (an enzyme that converts fatty acids)* or metabolized into *Prostaglandin E1* by *cyclooxygenase (involved in the creation of inflammatory mediators)*, both of which have anti-inflammatory properties.

- *15-HETrE* can block the oxidation of **arachidonic acid (AA)** into *Leukotriene A4 (LTA4)* by inhibiting *5-lipoxygenase (5-LOX); 5-LOX* converts AA into *leukotrienes. Leukotrienes* are involved in inflammatory and allergic conditions as well as *vasoconstriction (constriction of blood vessels), LTA4* has been seen as the basis for the creation of pro-inflammatory *leukotrienes.*
- *Prostaglandin E1* has been seen as a vasodilator; it has also been seen to have *anti-thrombotic (anti-clotting)* effects by activating *Prostaglandin receptor; this* can lead to increased *cyclic Adenosine Monophosphate* levels in smooth muscles resulting in lower *Ca^{2+} (Calcium Ion)* levels within the cells allowing relaxation of blood vessels, since *Ca2+ is* involved in muscle contraction.

This mechanism can give GLA anti-blood clotting effects, which can promote cardiovascular health. **Borage Oil** or **Evening Primrose Oil** have been seen to have high levels of GLA.

- It's important to note that many people like to consume CLA for fat loss effects. Do not confuse GLA with **CLA**. GLA and CLA *have different structures and evidence; neither mechan*ism establishes reliable weight loss from GLA.

Proposed benefits (evidence varies)

- May aid in recovery
- Vasodilator
- Anti-inflammatory

Side Effects

- Low blood pressure
- Headache
- Nausea

Dosage

- Trials use product-specific amounts; total oil weight differs from GLA dose. No general health or **diabetic-neuropathy regimen is established.**

WARNING

Seek medical advice before use. GLA may affect those with **bleeding disorders** as GLA can affect clotting. GLA can lower blood pressure; this can lower blood pressure to unsafe levels if you have **low blood pressure.**

GPLC (Glycine propionyl-L-carnitine)

GPLC is a **Carnitine** molecule attached to the amino acid *glycine* and can be seen marketed as a general performance enhancer, capable of stimulating *Nitric Oxide Synthesis*, as well as having an antioxidant effect in the body. *Nitric Oxide (NO) is* extremely vital to health, as it is able to relax smooth muscles; e.g. heart, keep blood platelets from sticking to each other thereby increasing blood flow to skeletal muscles and reducing the risk of clots.

GPLC is believed to provide three athletic benefits with supplementation: one being enhanced *Nitric Oxide* levels due to the **Carnitine** molecule, lastly would be power enhancement due to the *propionyl* molecules as they may be converted into *succinate (which can re-enter the Krebs Cycle, reference the third bullet)* aiding in energy production in the *Krebs Cycle* through the creation of *thiols.*

- Carnitine is believed to be implicated as a possible antioxidant of *Nitric Oxide* as it has been seen to be able to stimulate *Glutathione* expression as increased glutathione levels in the blood can reduce *nitric oxide* breakdown. Carnitine is also believed to promote *Nitric Oxide* expression at a transcriptional level, possibly relating to *endothelial Nitric Oxide Synthase (involved in vasodilation).*
- It has also been suggested that Carnitine can reduce lactate build-up in muscle tissue due to rigorous exercise. However, studies to date have shown no effect of lactate levels in muscle tissue nor any alteration.
- The increased power benefit, however, seems to come from the *propionyl* molecules, not the reduced levels of lactate. *Propionyl* is a *propionic acid* that is involved in the *Krebs Cycle. Propionic acid,* the metabolites of various amino acids (valine, isoleucine, **methionine,** and threonine), fatty acids and the side chain of cholesterol are funneled to create Propionyl-CoA; Propionyl-CoA is converted by *propionyl-CoA carboxylase* into *methylmalonyl-CoA*, which is converted into *succinyl-CoA* by *methylmalonyl-CoA mutase*. *Succinyl-CoA* can then enter the *Krebs Cycle* as a *thiol. Succinyl-CoA* is stripped of *Coenzyme A (CoA),* breaking the *thiol* bond.

A *thiol* bond is an extremely high energy bond that happens to be *exergonic (dissipates to the surrounding area); the energy* released is then used for **ATP** and *GTP (GTP used for RNA synthesis).*

Benefits

- Enhance Nitric Oxide Synthesis
- May increase endurance
- Vasodilator
- May promote fat loss
- May aid recovery

Possible Side Effects

- Fluctuating blood pressure
- Stomach Cramps
- Headache

Dosage

- There is no set dosage; adhere to label directions. A possible dose is 2-4.5 grams prior to work out.

WARNING

Please consult with a medical professional prior to administering GPLC. L-carnitine has been seen to interact with thyroid hormones; if you are diagnosed with **thyroid problems,** GPLC may make symptoms worse or treatments less effective. L-carnitine has also been linked to an increase in seizures in those already prone to **seizures**. If you are under **blood thinners/blood pressure** Carnitine can affect blood pressure due to its effect with nitric oxide which would also affect platelet adhesion and **clotting**.

Gamma Oryzanol (γ-Oryzanol)

γ-Oryzanol is a compound of four molecules that build off a *ferulic acid* backbone, which are found in Rice Bran Oil; it makes up .1-2% of the oil sometimes less, so the same benefit will not be observed through regular oil consumption. It can be found marketed as a supplement to increase strength, lower cholesterol, and improve skin aesthetics. While there are statements out there stating that Gamma Oryzanol is a testosterone booster, mechanistically speaking, there is no reason as to how it would accomplish this, also, there are no studies showing an increase in free testosterone levels. A resistance-training trial did not establish improved strength or body composition; it is not a proven anabolic supplement.

It is believed that the main mechanism seen in γ-Oryzanol comes from its antioxidant capabilities and its modifications in lipid metabolism. Various *ferulates* have shown antioxidant benefits as they can scavenge free radicals responsible for oxidative stress and cellular damage like *Reactive Oxygen Species (ROS). ROS* is used by the body for various processes and signaling. When *ROS* is produced to excess especially in high-fat diets; free radical damage in the body is increased leading to *lipid peroxidation. Lipid peroxidation* is oxidative damage to cells, in which free radicals compromise the integrity of lipids in cell membranes. *Ferulates* are also structurally similar to cholesterol so they may displace cholesterol which can lead to a reduction of *hepatic lipogenesis* reducing cholesterol synthesis.

- *Lipid peroxidation* can increase the risk of *mitochondrial dysfunction (metabolic dysfunction)* due to *lipotoxicity (fat metabolites stored in non-fat tissue leading to cell dysfunction and death).*

γ-Oryzanol and its *ferulates* have been seen to promote *adiponectin (a hormone involved in fatty acid breakdown and glucose regulation)* expression. This expression can then activate *5' Adenosine Monophosphate-Activated Protein Kinase (AMPK)* in *adipocytes (fat cells). AMPK* activates fatty acids or *glucose* usage when cellular energy is low to maintain optimal cellular energy levels; this increased uptake of fatty acids and triglycerides can improve glucose tolerance and cholesterol reduction.

- *Ferulic acid* has been seen to reduce *thromboxane* B_2 production. *Thromboxane 2* can facilitate platelet aggregation as well as be a vasoconstrictor *(constriction of blood vessels and vascular muscle cells).* This can promote *Endothelial NOS* potentially reducing clots.

Ferulic Acid can also reduce oxidative stress on the skin when applied topically, thereby reducing *thymine dimers (these can cause cancer)* on the skin.

Proposed benefits (evidence varies)

- May improve strength
- Improved skin appearance
- Antioxidant
- Vasodilator
- Anti-inflammatory
- Antidiabetic

Possible Side Effects

- Dry mouth
- Low blood pressure
- Irritability

Dosage

- 500 to 600 mg of gamma oryzanol a day

***WARNING**

Seek medical advice before use. Γ-Oryzanol can affect **hypothyroidism**, as it can lower thyroid function. γ-Oryzanol can also affect **blood pressure/clotting,** affecting those with blood pressure and clotting problems. γ-Oryzanol has also been seen to affect glucose levels; this may affect **diabetics**; insulin medication may need to be modified.

Garcinia cambogia

Garcinia cambogia comes from the *Clusiaceae* family and the fruit has been described as a green pumpkin-like shaped fruit. Some extracts have been known to go under the name of brindleberry or malabar tamarind. It has been used throughout the last few decades to enhance food satisfaction and as a weight loss supplement. Prior to continuing further, there's an important disclaimer: individuals with possible ***Liver problems****, in particular, Cytochrome P450 deficiency, there is concern that Garcinia Cambogia may cause hepatotoxicity due to previous/current liver damage or the liver's incapability to break Garcinia down.* Serious liver injury has been reported with Garcinia-containing products, including in people without known prior disease; causation can be difficult to separate in mixtures.

The active ingredient in Garcinia cambogia is *hydroxycitric acid (HCA).* There has been much controversy over its efficacy. Current evidence indicates little or no meaningful weight-loss benefit; lack of efficacy cannot simply be attributed to improper use. Partially, there has been no controlled way of consuming it. Many companies and individuals claim that the addition of potassium and magnesium protects *HCA*, keeping it from becoming inactive, allowing the body to absorb it better.

It has also been seen to satisfy one's consumption of the meal to a greater extent, meaning one would cut down on the amount they ate due to *satiety (the feeling of being satisfied/full). The reason for this* is believed to be because *HCA* might be able to inhibit the *reuptake (the absorption of neurotransmitter by its receptor)* of *serotonin*. Generally, this effect is observed when consumed prior to a meal.

- *HCA* has been seen to block the enzyme *ATP-citrate-lysase*, responsible for turning starches and sugars to fat as activated by insulin. Meaning fat creation from carbs will be reduced. However, this mechanism has not been fully verified in humans as it was only noted in rodents.

Other marketed fat burners also have limited or modest evidence; no supplement replaces an effective weight-management plan. The full mechanism and effects of Garcinia cambogia are not fully known.

Proposed benefits (evidence varies)

- May aid in fat loss
- May promote meal satisfaction

Possible Side Effects

- Gastrointestinal discomfort
- Nausea
- Headache

Dosage

- Dosage is 500 mg 3x a day, combined with **Potassium** and **Magnesium** 30-60 minutes before a meal.

WARNING

Seek medical advice before use. Garcinia cambogia has been known to cause gastrointestinal discomfort; if you have any **digestive** problems, please consult a doctor. The fact that *hydroxycitric acid* may inhibit sugar and carbohydrate conversion into fatty acids might affect insulin levels. Please consult with a doctor if you are **diabetic** as it could affect blood sugar levels and medication might need to be adjusted.

Garlic *(Allium sativum)*

Garlic comes from the *Allium* family where onions and chives can be found and has been one of the most universally recognized health foods and medicinal supplements. It is native to Asia and parts of Iran, while China is the main global producer of Garlic. It has been used in traditional medicine to aid and reduce the risk of cardiovascular diseases while acting as a strong antioxidant.

The main active component in garlic is *alliin.* When *alliin* breaks down, *allicin* is formed which can then be broken down even further into *hydrogen sulfide.* For general health benefits, it is recommended to eat raw garlic. As *allicin* becomes activated when garlic is broken or crushed, shortly after it is converted into its reduced form such as *hydrogen sulfide (sulfide=garlic smell*), making it extremely unstable.

- *Hydrogen sulfide* can act as a vasodilator by stimulating *nitric oxide (NO, involved in vasodilation),* and directly relaxing blood vessels and their *endothelium* keeping them from sticking to each other; it is believed to accomplish this by breaking down *S-nitroglutathione (the reservoir of NO),* thereby increasing *NO* levels. This can aid high blood pressure diseases.

Another bioactive molecule of *allicin* is *thiacremonone* which appears to inhibit *NF-kB (where the stimuli for inflammation signals start)* by oxidizing **cysteine** proteins needed for *p50 subunit (recruits the needed molecules to activate NF-kB)* activity within *NF-kB.* This can result in a reduction of inflammation in the body. The bioactive compounds *diallyl disulfide* and *trisulfide* in garlic may also be able to protect the body from heavy metals, preventing the body from storing heavy metals in the organs. *Allylmercaptan* has been seen to inhibit *HMG CoA (the target of statins),* an enzyme involved in the production of cholesterol; inhibition of this enzyme increases cholesterol receptors in the liver; this results in the breakdown of cholesterol mainly *LDL,* lowering levels of cholesterol in the body.

- Garlic's bioactive compounds have also been seen to inhibit *lipogenesis (creation of fat cells)* and increase *brown fat* production. Regular fat (*white fat*) stores calories while *brown fat* is full of mitochondria and is used to burn energy and produce heat.

Benefits

- Vasodilator
- May reduce cholesterol
- Antioxidant
- Detox aid
- May increase energy
- Anti-fungal
- Anti-bacterial
- Fat loss aid
- Anti-parasitic

Possible Side Effects

- "Garlic" breath
- Heartburn
- "Garlic" body odor
- Diarrhea
- Gas

Dosages

- Adhere to label instructions; there are many extracts and forms of consuming garlic. The set dosage is based on the extract method.

WARNING

Please consult with a medical professional prior to administering Garlic. There is concern that garlic may affect those with **bleeding problems** as Garlic can reduce clotting. If you are diagnosed with **low blood pressure, garlic** can make blood pressure too low. It is important to note that garlic **CAN** become **TOXIC** at high doses.

Ginger Root *(Zingiber officinale)*

Ginger is part of the *Zingiberaceae* family where Turmeric can be found, and has been widely used in traditional Ayurvedic and Unani medicine to remedy a few ailments, such as nausea due to motion sickness through the consumption of ginger tea, as well as aid with colds, cramps and an upset stomach. It is native to Southeast Asia, and its stems have a unique rolled leaf visual characteristic (as in the stems look like they are leaves rolled one over the other). It can currently be seen marketed as an anti-inflammatory, antioxidant and a possible testosterone booster.

The active components are *gingerol* and *6-Shogaol; gingerol* has been seen to act as a powerful antioxidant, while *6-Shogaol* has been observed to be an anti-inflammatory as well as an antioxidant. <u>Although its full mechanism isn't fully known, the following has been observed.</u>

- Its inflammatory effect seems to be derived from the inhibition of *Tumor Necrosis Factor Alpha, Interleukin-1* and *Interleukin-8 cytokine (inflammatory messengers)* production; these messengers are proinflammatory and their production can lead to swelling.
- Ginger has also been observed to inhibit the pro-inflammatory pathway of *cyclooxygenase (involved in the creation of inflammatory messengers)* and *5-lipoxygenase (5-LOX). 5-LOX* converts **arachidonic acid (AA)** into *leukotriene A4. Leukotriene A4 is* involved in inflammatory and allergic responses as well as *vasoconstriction (constriction of blood vessels).*

Ginger's antioxidant effects seem to be derived from raising *Catalase* and *Superoxide Dismutase; these* enzymes are involved in breaking down free radicals like *Reactive Oxygen Species* which are known to cause cellular damage which can potentially alter DNA enough to cause cancer.

- *6-gingerol* has been seen to promote *Tumor Protein p53 Bcl-2-associated X (BAX)* protein production; BAX is involved in *apoptosis (programmed cell death)* activation, while *p53* has been implicated as a tumor suppressor.

Ginger root seems to be able to increase testosterone levels in the body, although the reasons as to why are still unknown; however it is believed to increase cholesterol content in the testicles as this is where *Leydig cells* are located at which synthesize cholesterol

<u>Benefits</u>

- May boost testosterone
- May be anti-carcinogenic
- Reduces nausea
- Osteoarthritis aid
- Antioxidant
- Anti-inflammatory

<u>Possible Side Effects</u>

- Gastrointestinal discomfort
- Heartburn
- Diarrhea

<u>Dosage</u>

- 1 to 3 grams of ginger extract
- 28 mL of ginger extract for boosting testosterone

<u>***WARNING***</u>

Please consult with a medical professional prior to administering Ginger root. There is concern that ginger root may affect those with **bleeding problems** as Ginger can affect clotting. Those with **low blood pressure**, ginger root can cause blood pressure to drop to unsafe levels. Ginger root can lower blood glucose levels which can affect diabetics; insulin medications may need to be modified.

Ginkgo Biloba

Ginkgo biloba is actually a tree from the *Ginkgoaceae family; this* tree has been known to live to upwards of 1,000 years and it's currently known as the only one of the *ginkgophyte* species that has not become extinct, although it is on the endangered species list. Its preservation can be mainly thanked towards Buddhist Monks as they cultivated it into their temple gardens, as they perceived it as a sacred tree. Currently it can be found marketed as a *nootropic (cognitive enhancer),* a neuronal antioxidant and as a cardiovascular health aid. Ginkgo has not been shown to prevent dementia; mechanisms do not establish reversal of Alzheimer's disease.

The main active components are *bilobalide* and *ginkgolides*. *Bilobalide* seems to have a protective effect in the mitochondria as it is able to maintain **ATP** levels which can support mitochondrial respiration reducing mitochondrial dysfunction. Ginkgo biloba extract has been seen to restore proper mitochondrial function from defects caused by *Amyloid Beta (Aβ). Aβ* has been implicated in disrupting mitochondrial respiration, causing an overproduction of *Reactive Oxygen Species*. This can result in oxidative damage to neurons, further down impairing ion function within the cell, causing a calcium influx which can result in excitotoxicity reducing neurotransmission and causing synapse loss. This has been observed as the early stages of Alzheimer's Disease.

- *Bilobalide* has been seen to increase neuron cell *proliferation (increase/multiplication),* which might be responsible for its nootropic effects by increasing *Brain-derived neurotrophic factor (BDNF),* which is responsible for neuron survival and the growth of new neurons and *synapses.*

Similar to **bacopa monnieri**, ginkgo biloba is able to increase blood flow to the brain; this may be the result of the *ginkgolides*. *Ginkgolides* seem to inhibit the *platelet-activating factor (PAF). PAF* is responsible for the constriction of blood, lung and platelet aggregation. By inhibiting *PAF*, blood flow can increase to the brain and may reduce damage brought on by stroke as well.

- This vasodilation effect also applies to the rest of the body.

Ginkgo biloba may also protect blood cells from *Low-density Lipoprotein (LDL)* oxidizing (this can cause oxidative damage to blood cells) as well as reduce *foam cells* through its antioxidant effects. *LDL* can attach/cause *macrophages (white blood cells)* to become foam cells; foam cells *accumula*te in artery walls, which can lead to *atherosclerotic* build-up.

Proposed benefits (evidence varies)

- May aid Alzheimer's Disease
- Aid neuron repair
- Antioxidant
- Vasodilator
- Nootropic

Possible Side Effects

- Upset stomach
- Constipation
- Dizziness
- Headache

Dosage

- 60 mg of ginkgo biloba extract (EGb-761 50:1 ratio) 3x times a day

WARNING

Seek medical advice before use. Ginkgo Biloba may affect those with **bleeding problems** as it can affect clotting. Ginkgo Biloba can affect blood glucose levels; **Diabetics' insulin** may need to be modified. If you are **highly allergic** to poison ivy/oak or cashew oil, Ginkgo Biloba may cause an allergic reaction as it contains Urushiol, the sap that causes the rash experienced with poison ivy.

Ginseng (Panax Ginseng)

Ginseng refers to the root of the plants in the *Panax genus (hierarchy of classification, one step below family).* Panax Ginseng is the most commonly known and researched ginseng. Native to Asia, predominantly found in China and Korea, it has been used in traditional Chinese medicine to promote virility and boost general well-being. Currently, it can be seen marketed as an athletic enhancer, testosterone booster, an immune and cognition booster. Panax Ginseng is nearing the endangered species list mainly because demand has surpassed cultivation levels; this has led to the use of "fake ginseng" herbs such as *Codonopsis pilosula* as filler products which can affect the body differently. *No consistent testosterone or athletic-performance benefit is established; results depend on the preparation and condition.*

The bioactive compound found in Panax Ginseng is *ginsenoside* and it seems to affect and target various areas in the body. This is due to the fact that *ginsenosides* are a classification for over 100 *ginsenosides* all classified down to 4 subtypes: *Protopanaxadiol, Protopanaxatriol Oleanolic,* and *Ocotillol.*

- *Protopanaxadiol* has been implicated with reducing *Reactive Oxygen Species (free-radicals that cause cellular damage)* oxidative stress due to *influenza.* The reduction can lead to an anti-inflammatory effect, inherently reducing DNA damage and *apoptosis (programmed cell death).*
- *Protopanaxatriol* has been seen to stimulate and sedate the Central Nervous system. It can also stimulate *Type 1* and *2 T helper* cells, as it's been observed that *interferon-gamma (involved in innate/adaptive immunity)* and *interleukin-5 (involved in antibody and white cell production)* production was increased; this immune enhancement can support *pathogen* eradication.
- *Oleanolic* has been seen to inhibit *5α-reductase,* which may be the reason for its testosterone boosting benefits. *Enzyme 5α-reductase* is responsible for turning *testosterone* into *dihydrotestosterone (DHT).* In theory, the inhibition of DHT conversion would create an accumulation of testosterone in the body which could enhance muscle development.
- *Ocotillol is believed to* can promote neuronal activity as it has been seen to have excitatory effects; increasing *glutamate transmission* and *release* in *mitral cells (neuronal cells that take signals from odor and transfer them to various parts in the brain).* However, current cognitive enhancing benefits are currently not fully known.

Proposed benefits (evidence varies)

- May boost testosterone
- Promote virility
- Nootropic
- Anti-inflammatory
- Antioxidant
- Immune booster

Possible Side Effects

- Diarrhea
- Mood changes
- Itching
- Insomnia
- Increased hear rate
- Loss of appetite

Dosage

- 200-400 mg of a standardized extract of 3% *Ginsenosides* once a day

WARNING

Seek medical advice before use. Panax Ginseng can affect those with **immune problems** as Panax has been known to stimulate immune function. Those with **low blood pressure, Panax** can cause blood pressure to become dangerously low. **Hormone-sensitive** conditions such as **ovarian** or **uterine cancer** or **endometriosis/uterine fibroids, the** *ginsenosides* can act like estrogen worsening symptoms. Review glucose-lowering medicines and insomnia risk; avoid use during pregnancy without specialist advice.

Glucomannan

Glucomannan is a water-soluble fiber supplement that's extracted from the *Elephant Foot Yam (Amorphophallus paeoniifolius from the Araceae family)*, which is native to Asia. It has generally been used as food additives for emulsifying or thickening, but currently, it has gained ground as a weight loss supplement where the benefit has mainly been observed in overweight/obese individuals. It can also be seen marketed as a cholesterol and diabetes aid.

Glucomannan is made up of a *polysaccharide (carbohydrate)* composed of *beta-D-mannose* and *beta-D-glucose; the* linkage between these two carbohydrates cannot be broken down by salivary enzymes nor by other enzymes found in the digestive tract, allowing it to pass straight through the colon unchanged, where bacteria there can break it down and ferment it. It is also able to retain 50 times its weight in water making it extremely viscous.

- The main reason weight loss is achieved through glucomannan is that it is able to expand in the gastrointestinal tract, thereby promoting feelings of fullness and satiety.
    - This can lead to slow absorption of food, which can reduce appetite.
    - Through the same mechanism, it is able to reduce glucose uptake, thereby decreasing insulin spikes.
- In a similar manner, it is able to reduce *Low-density Lipoproteins (LDL)* by cholesterol fecal excretion; this is due to a reduction in the absorption of cholesterol in the intestinal tract.
    - In this sense, this is how it aids in constipation.

Theoretically, this is how it would work, although there is a handful of studies promoting effectivity while others were seeing no change, leading to contradicting evidence. No conclusive data has been found. A good portion of results stated no appetite curb or hunger reduction, but noted an increase in bloating, belching and stomach fullness. There is a lack of data stating what this product is meant to do. *It is also important to note that much of the data that had made these claims is extremely outdated (20 years or so)*.

Proposed benefits (evidence varies)

- May lower blood glucose levels
- May lower cholesterol
- May aid in weight loss

Possible Side Effects

- Diarrhea
- Abdominal discomfort
- Gas
- Bloating

Dosage

- Follow the specific product's studied dose and take with plenty of water; dry products can cause choking or obstruction.

WARNING

Seek medical advice before use. Monitor glucose with diabetes medicines. Swelling can cause choking or esophageal or intestinal obstruction, especially with **structural abnormalities** or swallow**ing difficulty. Take with adequate** fluid; do not swallow dry powder. Weight-loss results are inconsistent.

Glucosamine

Glucosamine is an amino sugar that is naturally found and created within the body to help repair cartilage and joints. Glucosamine has long been used for joint health and to reduce inflammation in the joints as well. With an added note that *glucosamine sulfate* can more readily enter the body, compared to *glucosamine hydrochloride*, the reason for this is not yet known. The degree of effect and efficacy in joint health and cartilage degradation attenuation is still up for debate. Nonetheless, long-term supplementation has been reported to provide some aid and comfort. Clinical results vary by formulation and joint condition; cartilage regrowth and rheumatoid-arthritis treatment are not established.

- There is no immediate glucosamine transport in the intestines. Since glucose has a similar structure it is believed that glucosamine seems to use that transport too; further down the intestines it is also believed that bacteria can metabolize glucosamine.

It was believed that glucosamine would enter the bloodstream and would then be taken up by *chondrocytes (cartilage cells)* which would in turn increase *proteoglycan synthesis (this and collagen make up cartilage).* This holds some merit as *proteoglycans* are made up of *glycosaminoglycans* which happen to have *N-acetylglucosamine* as part of its structure. This mechanism was further believed since glucosamine was seen to reduce *CTX-II (C-telopeptide of type II collagen),* a marker released during cartilage *resorption (break down).* However currently its mechanism is believed to be due to its anti-inflammatory effect, reducing the rate of **collagen** degradation due to inflammatory damage.

- It is believed to have efficacy through the inhibition of *interleukin-1β (IL-1β, a pro-inflammatory signaler); this* can further reduce *matrix metalloproteinase 2* & *matrix metallopeptidase 9* expression which are involved in the breakdown of *extracellular matrix* like the ones found in collagen. Additionally, *IL-1β* production has been linked with *cyclooxygenase-2 (COX-2) expression; COX-2* has been implicated in the aggravation of cartilage inflammation.

For those suffering from joint health problems, there are a few synergistic compounds that build on each other's benefit. **Chondroitin** and glucosamine have been known to promote *collagen synthesis in* ligaments, *chondrocytes (cartilage cells)* and *tenocytes (tendon cells), as well as inhibit* the breakdown of *collagen.* **Collagen** protein has also been seen to have synergistic effects with these two compounds.

Proposed benefits (evidence varies)

- Rheumatoid arthritis aid
- Promote joint mobility
- Osteoarthritis aid
- Reduce cartilage wear

Possible Side Effects

- Nausea
- Gas
- Heartburn
- Diarrhea

Dosage

- 300-500 mg 3x a day for general joint health
- 2-3 grams a day for high impact sport athletes

WARNING

Seek medical advice before use. Glucosamine may affect **diabetics** since it can raise blood sugar levels, as glucosamine has a similar structure to that of glucose. If you are **allergic** to **shellfish,** glucosamine might cause an allergic reaction as a major portion of glucosamine is derived from **shellfish**. Glucosamine can increase warfarin's effect; review use with the prescriber.

Glucuronolactone (D-Glucurono-y-lactone DGL)

Glucuronolactone can be found naturally occurring within the body, and since the creation of energy drinks, it can be commonly found in just about any major producers' energy drinks. For its widespread usage, however, there isn't much-known data or information on its benefits and effects. DGL is a metabolite of glucose and excreted by the body as sugar compounds, *xylitol, and xylulose.* DGL has been touted to boost the body's immune system all while promoting an anti-cancerous environment within the body. There aren't many studies done on DGL. Although no short/long term toxicity has been reported at recommended dosages, some anecdotal reports have stated positive reviews.

DGL has also been noted to have antioxidant effects as it is able to reduce oxidative damage to *lipoproteins. Lipoproteins* are important as they help fats be transported into cells for use since fats are insoluble in water; they accomplish this by making the *hydrophobic (water-hating)* fats into *hydrophilic (water-loving)* molecules. This may provide a protective effect on blood vessels, as a type of *lipoprotein (chylomicrons),* interacts with the lining of blood vessels, thereby also inhibiting platelet aggregation. This is mainly believed to be due to the *chylomicrons* reducing *thrombin (an enzyme involved in blood clotting)* induced platelet aggregation.

- DGL may have liver-protective properties; this might be a result of DGL being a sequestrant; sequestrants are capable of chelation; chelation is the bonding of toxic metal ions to a molecule making them easier for the body to remove, thereby preventing damage to cells.
- DGL's anti-cancerous properties seem to come from its inhibition of *β-glucuronidase; β-glucuronidase* are enzymes that break down carbohydrates and have been seen to be overactive in various cancers (cancers tend to use *glycogen* for energy instead of oxygen since they are anaerobic) seen in individuals when compared to healthy individuals.

DGL also regulates *glycogen (stored energy; refer to ATP for an in-depth explanation) synthesis; certain* drugs have been known to deplete DGL; this can cause *glycogen* levels to drop reducing performance. DGL supplementation may promote *glycogen* synthesis, allowing for enhanced physical performance.
In animals, it has been seen to increase performance after continuous supplementation, although this has not been clinically tested in humans.

Benefits

- Anti-cancerous
- Detox aid
- Antioxidant
- May improve endurance

Possible Side Effects

There are currently no known side effects for lack of data/anecdotal reports

Dosage

- 10-60 mg a day/it has synergistic properties with **resveratrol**

WARNING

Please consult with a medical professional prior to administering Glucuronolactone. There is concern that glucuronolactone may affect glucose levels in **diabetics** as it is a type of glucose; medication may need to be modified. If you are diagnosed with any sort of **liver dysfunction**, please consult with your doctor as glucuronolactone is believed to affect the liver. It is currently unknown if it may aid or worsen symptoms. For those with hypertensive conditions, Glucuronolactone is generally paired with caffeine which can raise systolic blood pressure.

Glutamine

Glutamine is an amino acid that plays a role in various functions within the body. This amino acid can be found in meats, eggs, beans, cabbage, kale as well as protein powders. It is also the favored fuel source for the immune system and the intestines. Glutamine is a large fraction of the muscle free-amino-acid pool, not 60% of muscle tissue, and it is the main transporter of *nitrogen* in muscle tissue. It currently can be seen mainly marketed towards athletes as a support and aid in exercise recovery; at times it can also be found as a beneficial supplement for vegans or vegetarians. Its physiological role does not establish added muscle gain or exercise benefit in healthy athletes; routine training does not itself prove deficiency.

Glutamine has been shown to promote muscle *anabolism (constructive metabolism)* due to a surplus on glutamine. Glutamine has been seen to have protective effects on **Leucine** by reducing the oxidation of leucine. L-Leucine plays a role in activating the amino acid-dependent *Rag GTPase heterodimers* in *mTORc1 (plays a role in protein synthesis).* Leucine availability helps regulate this pathway, alongside other amino acids and signals. When these levels are met the *heterodimers* are switched to "on"/active, starting a chain reaction in *mTORc1* that activates protein synthesis. The pathway of this mechanism can promote an *anabolic* environment that favors muscle tissue growth or general tissue growth. This mechanism is beneficial for enhancing recovery and in damaged tissue following injury or surgery.

- L-Leucine has been seen to increase *mitochondria* biogenesis by *SIRT1 activation; through this* an increase in *mitochondria* in muscle cells can increase resistance to exercise-induced fatigue.

Other systems in the body need glutamine for fuel like the immune system or the intestines, meaning that glutamine use in the body is prioritized; this means that skeletal muscle is not first or second.

- Through glutamine supplementation, the immune system is able to stay boosted and optimal, since glutamine can and is used more efficiently by *leukocytes (white blood cells)* than glucose which can enhance their growth. Supplementation can provide fuel for use in other systems which can prevent muscle tissue *catabolism (breakdown).*

Glutamine has been implicated in the regulation of intestinal *tight junction barriers (selective semi-permeable barriers in the intestines)* aiding intestinal integrity.

<u>Proposed benefits (evidence varies)</u>

- Maintain an anabolic environment
- Recovery aid
- Boost immune system

<u>Possible Side Effects</u>

- Nausea
- Frequent bowel movements
- Headache
- Gas
- Vomiting
- Rash

<u>Dosage</u>

- No established athletic benefit supports a routine 0.75 g/kg/day regimen. Clinical high-dose use is purpose-specific and requires supervision.

WARNING

Seek medical advice before use. Glutamine may worsen symptoms of people with **cirrhosis** as well as **hepatic encephalopathy**. If you are **allergic/sensitive** to **monosodium glutamate (MSG)** you might have the same reaction from glutamine as the body converts it into glutamate. If you are prone to **seizures,** glutamate may increase seizure frequency.

Glycerol *(1,2,3-propanetriol)*

Glycerol, also known as *glycerin* or *glycerine,* is a *polyol (a type of sugar alcohol)* that has been used as a sweetener by the food industry. As of late, Glycerol has gained a reputation and can also be found marketed in the supplement industry for having similar hydrating effects like **creatine** and a possible endurance booster.

Glycerol can be naturally produced in the body through *glyceroneogenesis* with the use of **glutamine, alanine,** and **pyruvate**. Glycerol can be used by the body for *gluconeogenesis (creation of sugars like glucose)* in times of low energy consumption/nutrition such as fasting, starvation or during bouts of high intensity. During prolonged periods of starvation, glycerol becomes the only source used in *gluconeogenesis.*

General hydration is a basic fundamental element in physical performance; lower hydration results in lower performance, higher hydration in the body leads to optimal performance. Glycerol is able to induce "hyperhydration" in cells. Glycerol supplementation accumulates in body fluids, except within the eyes or the brain; this can cause *osmotic (equilibrium of water/fluids)* pressure and water volume within the body to increase, hence the given effect in hyperhydration.

- When consumed, glycerin is absorbed by the small intestine; there it's evenly distributed throughout the body.
- Thereafter water absorption in the *nephron (regulates water/electrolyte concentration)* is enhanced. Through this, "hyperhydration" of cells is promoted through an osmotic gradient, in short balancing water/electrolyte percentages in targeted areas.

Glycerol can cause weight gain; however, this side effect is believed to be due to its water retention effects; this mechanism has also been implicated in enhancing skin aesthetics.

In simpler terms, glycerol is able to induce fluid retention in the body and possibly promote endurance through an added energy source. There's currently not much data in Glycerol outside of this, due to its low toxicity levels (toxic levels near 5g per kg/2.2 lbs of body weight), it is deemed generally safe.

Benefits

- May increase power
- Cell "Hyperhydration"
- May Increase Endurance
- Enhance skin tenacity

Possible Side Effects

- Headaches
- Dizziness
- Nausea
- Vomiting
- Thirst
- Bloating

Dosage

- 1 gram per 2.2 lbs/1 kg per body weight along with 1.5 L of water 2 hours prior to an event

WARNING

Please consult with a medical professional prior to administering Glycerol. Glycerol should never be introduced through IV (Intravenously) as it can damage blood cells. Glycerol can affect those with **Glycerol Kinase Deficiency** as it can increase glycerol blood and urine to excessive levels.

Goji Berry or Wolfberry *(Lycium barbarum/chinense)*

Goji berries can be referred to as *Lycium fruits* since they are derived from two related plants, *Lycium chinense* and *Lycium barbarum*. Both plants belong to the *Solanaceae* or *Nightshade* family where potatoes and eggplants can be found. Goji berries have been used in traditional Chinese medicine for therapeutic benefits. They have gained popularity amongst various other berries these last few years for their claimed antioxidant abilities and can be seen currently marketed as a superfood.

The main benefit that is seen with goji berries is a mixture of its entire makeup, *polysaccharides (large/complex carbohydrates)*, *carotenoids*, phenolics, and its macro/micro-nutrients.

- It can protect the skin from UV damage by reducing the inflammatory response as well as inhibit the formation of blisters. It is able to resist *lipid peroxidation* that is also induced by UV light. *Lipid* Peroxidation is oxidative damage to cells where free radicals compromise the integrity of lipids in cell membranes and will continue as a chain reaction of free radicals.
    - The *polysaccharides* have been implicated in enhancing the immune system, also having *apoptotic (programmed cell death)* and *anti-proliferative (multiplication)* effect on cancer cells, although this mechanism is not fully understood.
- The *carotenoid* found in Goji berries, *Zeaxanthin (a bioactive pigment)*, has also been implicated in preventing the onset of *Age-related Macular Degeneration (AMD)*, by protecting the eye from *hypopigmentation (the loss of color)* as well as *drusen (yellow fatty protein deposits behind the retina)* accumulation in the eye, mainly as it's the foremost *carotenoid* in the *central macula (pigmented area responsible for high-resolution color vision)* which can increase *macular* pigment density.

Its antioxidant effects seem to be derived from *zeaxanthin* and *quercetin*, amongst other phenolic compounds, mainly via protection and prevention of oxidation from compounds such as *Reactive Oxygen Species (ROS, is a reactive Oxygen that can cause free radicals and damage to the body and cells)* as they can sequester free radicals while also promoting *Catalase* activity through *mRNA*. *Catalase is an enzyme* involved in breaking down free radicals like *ROS* which are known to cause cellular damage.

Benefits

- Immune system boost
- Support Eye Health
- Antioxidant

Possible Side Effects

- Gastrointestinal discomfort
- Low blood pressure
- Dizziness

Dosage

- If taking an extract adhere to label directions; potency can differ from brand to brand. Normal berry consumption is 30 grams of goji berries.

WARNING

Please consult with a medical professional prior to consuming Goji Berries. There is concern that Goji Berries may increase the risk of bleeding if on **blood thinners**. Goji berries can affect **diabetics** as it has been seen to reduce blood sugar levels; medication may need to be modified. If you are taking **blood pressure medication, goji** berries have been known to have anti-hypertensive effects, lowering blood pressure to unsafe levels.

<u>Goldenseal (*Hydrastis canadensis*)</u>

Goldenseal, also known as orangeroot and yellow puccoon, is from the *Ranunculaceae* family where buttercups can be found, and is native to North America. It has traditional antimicrobial uses. Currently it can be seen marketed in a similar fashion combined with **Echinacea** for cold remedies. It can also be seen advertised to aid gastrointestinal symptoms as well as having anti-parasitic properties. Current evidence does not support goldenseal for the advertised conditions; isolated berberine studies cannot be transferred to the whole herb.

Its main bioactive components are *berberine*, *hydrastine*, and *canadine* alkaloids. <u>There are, however, some unknown components</u> in this plant that act as *efflux pump inhibitors (EPI)*

- *EPI* stops the bacteria's ability to remove "toxins" from inside itself, allowing antibiotics/antimicrobials to become effective against antibiotic-resistant bacteria.

These unknown compounds act synergistically with the alkaloids found in Goldenseal, especially since *berberine* is an antimicrobial. Meaning it can promote a reduction of symptoms seen as a result of the oxidative response, or breakdown, of Gram-negative bacteria, in which *Lipopolysaccharide (LPS), an endotoxin* that activates *Toll Like Receptor 4 (involved in inflammatory signals),* is released which can cause system-wide inflammatory responses.

Berberine is the most abundantly found alkaloid in the plant, followed after by *hydrastine* then *canadine.* These alkaloids are powerful antimicrobials; the majority of Goldenseal extracts generally come from the root (also known as *rhizomes*), as that's where the highest concentration of alkaloids are located. It is important to note, however, *Berberine* has shown benefit in diabetics as well in small glucose-control trials, but this does not establish equivalence to *metformin or justify replacing prescribed treatment.* <u>*Consuming berberine in diabetics could cause interactions with any current medicine.*</u>

<u>Proposed benefits (evidence varies)</u>

- Anti-bacterial
- Anti-inflammatory
- Antidiabetic

<u>Possible Side Effects</u>

- Gastrointestinal discomfort
- Dizziness
- Vomiting
- Nausea
- Diarrhea

<u>Dosage</u>

- No established effective general dose exists for goldenseal; extract concentrations and interaction risks vary.
- Berberine and goldenseal are not interchangeable products; medicine review is needed before either is used.

<u>***WARNING***</u>

Seek medical advice before use. Goldenseal may interact with particular medications such as *warfarin, thiopental, tolbutamide,* and macrolide antibiotics like *azithromycin.* Goldenseal can inhibit drug-metabolizing enzymes and change medicine exposure; interactions are not simply displacement from an intended target. Review all medicines with the prescriber. There is concern that Goldenseal can also affect those with **Liver problems** as it has been observed to cause damage to liver cells (HepG2 cells); however, the mechanism isn't fully known yet.

Gotu Kola *(Centella asiatica)*

Gotu Kola can be known by many names, but one of the names it should not be confused with is Kola Nut. Kola Nut is not part of the same family and it has different effects on the body. Gotu Kola has been used in traditional Ayurveda medicine. Currently, it has been marketed as a nootropic (cognitive enhancement), a cardiovascular health aid and to help with recovery after an injury. Neurological and cancer mechanisms are mainly preclinical; reliable broad cognitive or disease-treatment benefits are not established.

The main active components in Gotu Kola for recovery and cardiovascular health are the saponins: *asiaticoside, madasiatic acid,* and *madecassoside.*

- These saponins are able to enhance recovery from injuries by stimulating *collagen synthesis (creation of connective tissues and extracellular space)* while inhibiting the inflammatory response in the area; inflammation can lead to scar tissue through *hypertrophy (enlargement of cell size); this* can reduce elasticity and tensile strength in the area.
- They have also been seen to aid cardiovascular health by maintaining the integrity of the *vascular intima (innermost membrane); this* effect can apply to the entire vascular system. It is believed that this is due to the reduced activity of the enzymes involved in *mucopolysaccharide metabolism (involved in the breakdown of connective tissues)* within the vascular walls.

The active molecules for neurological enhancement are *brahmoside* and *brahminoside.*

- These components tend to have a sedative effect that can help reduce the physical effects of anxiety and stress, by binding to *cholecystokinin receptors. Cholecystokinin* is a hormone believed to modulate memory, anxiety, and memory, hence its purported intelligence enhancing effect.

The extract has also shown neuroprotective effects by protecting neurons from oxidative damage, in particular, *glutamate*-induced death, mainly through an increase in the antioxidants *Superoxide dismutase* and *catalase.*

- These extracts may have the ability to promote neuronal growth, through *dendritic arborization (where new dendrite branches are created); however, data* is still lacking on the mechanisms behind this and has been mainly observed in mice.

Proposed benefits (evidence varies)

- Increased injury recovery
- Cardiovascular health aid
- Nootropic

Possible Side Effects

- Nausea
- Upset stomach
- Drowsiness

Dosage

- 35-48 mg per 2.2 lbs or 1 kg of body weight

WARNING

Seek medical advice before use. Gotu Kola may affect those with sensitive livers. Liver injury has been reported; it should not be assumed to occur only through interactions or to spare hepatic metabolism. If you are diagnosed with **liver disease,** Gotu Kola might make symptoms worse or cause complications. If you will be or are on any type of sedative, Gotu Kola may enhance the sedation effects.

Grape Seed *(Vitis vinifera)*

Grape seed extract comes from the berries (surprise—grapes are berries) of the plant *Vitis vinifera* which is just a fancier way of addressing the common grape plant. Grape seed extracts' most potent benefit is its ability to modulate (control/modify) the body's own antioxidant enzymes.

Generally, *Reactive Oxygen Species (ROS)* are used by the body to signal the immune system and to keep a balance in the body. This system is controlled by antioxidant enzymes. When *ROS* is produced to excess, that's when you see free radical damage in the body. Grape seed extract is able to combat these effects with its antioxidant activity by promoting **glutathione** levels within the body.

- Additionally, the *phenolic (antioxidant group)* compounds in grape seed oil, are able to find *ROS (like hydrogen peroxide)* and inhibit their *lipid peroxidation.*
    - *Lipid peroxidation* is oxidative damage to cells where free radicals compromise the integrity of lipids in cell membranes and will continue as a chain reaction of free radicals.
    - Some of the *phenolic* compounds can have *cytotoxicity (toxic to cells)* towards tumor cells.
- *Polyphenols (antioxidant groups)* are able to inhibit **Arachidonic Acid (AA)** release, which is responsible for the creation of the inflammatory *leukotrienes* and *prostaglandins*. These molecules are responsible for inflammatory signals as well as promoting the adhesion of platelets, which could in turn negatively affect cardiovascular health.

Its high antioxidant profile has also been attributed to Grape Seed's chelation*; chelation* is the bonding of toxic metal ions to a molecule, making them easier for the body to remove.

The decrease in *lipid peroxidation (where free radicals steal electrons from lipids, compromising the cell membrane)* of *Low-Density Lipoprotein (LDL)* can result in a decreased risk of *foam cells* which can potentially aid cardiovascular disease synergistically with its anti-inflammatory properties.

- *LDL* can attach to *macrophages (white blood cells),* causing them to become *foam cells. Foam cells accumula*te in artery walls, which can lead to *atherosclerotic* build-up. (Or plaque build-up in the arteries for short.)

Benefits

- Enhance immune system
- Antioxidant
- Anti-inflammatory
- Aid cardiovascular health

Possible Side Effects

- Headache
- Itchy scalp
- Nausea
- Dizziness

Dosage

Grape seed extract can vary by potency and extraction method. Adhere to label instructions.

WARNING

Seek medical advice before use. Grape Seed may increase bleeding in people with **bleeding** problems or are on **blood thinners**. Those with **low blood pressure, Grape** Seed Extract can lower your blood pressure to dangerous levels. If you are **allergic** to grapes, grape seed extract can cause an allergic reaction.

Green Coffee Extract

Green coffee extract is derived from unroasted coffee beans and is high in *chlorogenic acid* compared to roasted coffee beans as the heat does break these molecules down; it does contain **caffeine** as well. *Chlorogenic acid* has antioxidant effects considering it's a *phenolic* compound. This product can generally be seen advertised as a cardiovascular health product and more recently to enhance fat loss. <u>It is important to note that data behind its fat loss is inconclusive.</u> Additionally, Green Coffee Extract has moderate bioavailability and absorption.

One of the metabolites of *chlorogenic acid* is *ferulic acid,* which could be the reason for cardiovascular health. *Ferulic acid* can stimulate and protect *acetylcholine* which induces *Nitric Oxide* production. *Nitric Oxide* relaxes blood vessel's *endothelium (their skin),* leading to vasodilation thereby lowering blood pressure in the body. *Chlorogenic acid* seems to have synergistic properties with caffeine as well.

- *Chlorogenic acid* is able to reduce *triglyceride* build up in the liver, where caffeine is ineffective.
    - This may be due to its metabolite, *Ferulic* acid, as it is able to increase fatty acid oxidation in the liver, which could result in higher energy production as fatty acids are one of the fuel sources for **ATP**.
- Caffeine is able to reduce circulating *triglycerides* in the body; this could result in lower cholesterol, although there currently is no data to support this.

Chlorogenic acid is also able to reduce insulin spikes after a meal and may be able to reduce carbohydrate uptake if used long term. This is believed to be accomplished by reducing glucose uptake after meals by negatively affecting *sodium ion* dispersal, which ultimately reduces *enterocytes'* ability to uptake glucose in the body.

- It could possibly aid in glucose regulation, transporting glucose to skeletal muscles instead of being deposited in fatty tissue, additionally reducing glucose production in the body by inhibiting *glucose-6-phosphatase (an enzyme involved in glucose levels); this* may make it an antidiabetic.

This mechanism is what's believed to cause the fat loss effect; with a proper diet and exercise, it could be an effective means at weight loss, as a stand-alone product; however, its effects are moderate at best.

<u>Benefits</u>

- Aid fat loss
- Vasodilation
- Antioxidant

<u>Possible Side Effects</u>

- Insomnia
- Restlessness
- Increased breathing rate
- Increased heart rate
- Headache
- Anxiety
- Insomnia
- Nausea

<u>Dosage</u>

- 120-400 mg a day, effects are dose-dependent, assess caffeine.

<u>***WARNING***</u>

Please consult with a medical professional prior to administering Green Coffee Extract. There is concern that Green Coffee Extract may increase bleeding risk in those with **bleeding disorders** as well as affect insulin levels in **diabetics**, so medication may need to be modified. If you have **osteoporosis,** there is concern that high amounts of caffeine can flush the calcium out of the body, caffeine can also worsen diarrhea in those with **irritable bowel syndrome**. Those with **blood pressure** problems, Green Coffee Extract can lower blood pressure to dangerous levels.

Green Tea Extract *(Camellia sinensis)*

Green tea comes from the exact same leaves used to make black and oolong tea, the *Camellia sinensis* shrub, the only difference being the drying process. It has been associated as a beneficial additive for good health. Recently green tea extract has been marketed in various supplements and can be found at your typical nutrition store. Most benefits that are derived from Green Tea come from the water-soluble *polyphenols*. These compounds have been found to provide universal benefit to the majority of the body, from the cardiovascular system to the nervous system. Brewed tea and concentrated extracts differ: extract weight-loss effects are small, and liver injury has been reported with extracts.

One of the main active compounds in Green Tea is *Epigallocatechin gallate (EGCG),* an antioxidant whose measured activity depends on the assay; laboratory potency cannot be translated into a universal clinical advantage over Vitamin E.

- *EGCG* seems to have a nootropic effect as it is able to cross the blood-brain barrier, thereby increasing cognitive performance by promoting *proliferation (cell multiplication)* of *neural progenitor cells (neuron cells).* The observed mechanism behind this is an increased *Sonic Hedgehog Receptor (involved in the organization of the brain) mRNA transcription* in *neural progenitor cells* as well as increasing *GLI1 transcription (involved in Hedgehog signaling).*
- *EGCG* along with other *catechins* seems to reduce *LDL (Low-density lipoprotein)* cholesterol oxidation by inhibiting oxygen use in the oxidative reaction, as well as scavenging for metal ions; this can result in a reduction of *atherosclerosis (plaque build-up).*
- It is believed that *EGCG* in Green Tea may be the reason for vasodilation after consumption, as it is able to modulate *Nitric Oxide* by creating its own version of *free radicals; however, this* is still not fully understood.

Generally, Green Tea consumption may aid in suppressing the absorption of glucose that enters the bloodstream after a meal, but these effects are minimal at best and do not affect insulin production.

- Green Tea extract may aid at fat loss through *beta-adrenergic agonism,* when paired with **caffeine**, which acts as a thermogenic, *EGCG* can inhibit *catecholamines-o-methyl transferase (COMT),* which can inhibit caffeine's thermogenic effect by the activation of *noradrenaline.*

Proposed benefits (evidence varies)

- Nootropic
- Antioxidant
- Vasodilator
- Reduce cholesterol
- May aid fat loss

Possible Side Effects

- Upset stomach
- Dizziness
- Diarrhea
- Headache (from caffeine)
- Heartburn

Dosage

Potency and caffeine content may vary from brand to brand; please adhere to label directions.

WARNING

Seek medical advice before use. Green Tea may affect those with **anemia** as it can lower iron levels in the blood. The caffeine content might make **anxiety symptoms worse; for those** with **bleeding disorders**, green tea may increase the risk of bleeding. There is also concern in people with **osteoporosis** as the caffeine content can flush out calcium from the body.

Guarana *(Paullinia cupana)*

Guarana is part of the maple family, *Sapindaceae; it* is native to the Amazon and has been used by the people there as a stimulant. The seeds, about the same size of coffee beans, of this plant have been observed to have twice the amount of **caffeine** found in coffee beans. Currently it can be found marketed under energy drinks, fat loss supplements and even under some thermogenic products. The seed has been used in traditional Amazonian medicine.

Aside from its psycho-stimulatory effects brought on by caffeine, guarana seems to have benefits specific to its extract. <u>However, research is currently lacking and not all mechanisms or effects are known.</u>

- It has been seen to be a neuroprotector, being able to protect *dopaminergic cells (neurons that contain dopamine)* by increasing cell viability through stimulating *Catalase (an enzyme that protects cells from oxidative damage)* activity.
- Although the reasoning is not fully known, it has been seen that compounds in guarana may be able to enhance secondary memory like recalling tasks or recognition that's completely independent of caffeine.
- Guarana extract has been seen to have anxiolytic *(anti-anxiety) properties; this* i believed to be from a high *procyanidin* concentration, which seems to be associated with *serotonergic* and *dopaminergic* signaling modulation.

Guarana extracts may reduce platelet adhesion from *Arachidonic Acid* and *Adenosine Diphosphate* induced aggregation, as well as provide some antioxidant effects towards blood. Through this, it may aid *Nitric Oxide*, prolonging the vasodilation effect.

It is also important to not forget that guarana does contain caffeine, and care should be used to not encounter the side effects of caffeine. Please reference **caffeine** for an in-depth explanation of its effects and interactions, as that portion applies to guarana as well.

<u>Benefits</u>

- May aid cardiovascular health
- Nootropic
- Antioxidant

<u>Possible Side Effects</u>

The main side effects come from the caffeine content in guarana; guarana-independent side effects are:

- Stomach cramps

<u>Dosage</u>

- 75-100 mg a day, however, no optimal dose is known

WARNING

Please consult with a medical professional prior to administering Guarana. There is concern that Guarana may increase the risk of bleeding in those diagnosed with **bleeding disorders**. The caffeine content and the compounds in guarana may affect **diabetics** as it may complicate blood sugar control and increase symptom signs. If you are **allergic** to coffee beans, you may have an allergic reaction to guarana as it has similar chemical properties as coffee beans. For those on antidepressants, there is concern that Guarana can interfere with current medication as it has been known to affect dopamine and serotonin.

Guggulsterone

Guggulsterone is a *phytosteroid (not to be confused with anabolic steroids)* derived from the fragrant resin of the guggul tree also known as *Commiphora wightii* or *Commiphora mukul*. The guggul tree is part of the *Burseraceae* family where Frankincense (**Boswellia**) can be found; it has currently surfaced in the critically endangered list. It is native to India but can be found growing in parts of North Africa and Asia. It has been used in traditional Ayurvedic medicine. It is claimed to actively enhance thyroid functions, manage cholesterol and improve skin health.

Guggulsterone, although it has been used for thousands of years in ayurvedic medicine, is a fairly new supplement. There are different methods of extraction; the preferred method leaves you with a brown coffee like extract, all others are inferior to this. **Read the note below**.

- *Z-guggulsterone* can directly stimulate the thyroid by increasing *iodine* intake. It accomplishes this by enhancing *thyroid peroxidase (TPO); TPO* is an enzyme involved in the conversion of *iodine* into a usable form which is involved in the creation of the two Thyroid hormones, *thyroxine (responsible for metabolic rate)* and *triiodothyronine (involved in development, metabolism, body temperature, and heart rate)*. This can promote their release, which in turn might increase thermogenic/metabolic activity.
- Guggulsterone has been proposed to alter hepatic (liver) cholesterol by damping the *Farnesoid X (FXR) Receptor (involved in cholesterol regulation and HDL metabolism)*. This way, there's a decrease in cholesterol uptake by decreasing bile acid secretion, which means less uptake of cholesterol from food. A randomized trial did not lower LDL with guggulipid and found increases; receptor mechanisms do not establish cholesterol-lowering benefit.
- Guggulsterone is able to suppress the activation of *NF-kB (responsible for cell survival and signaling); reduction* in activation can lead to anti-inflammatory properties as well as reduce some cancer risks.

Proposed benefits (evidence varies)

- Lower cholesterol levels
- Anti-inflammatory
- Aid in fat loss

Possible Side Effects

- Gastrointestinal discomfort
- Nausea
- Headaches
- Rash
- Vomiting
- Belching

Dosage

- 30-60 mg 3x a day

WARNING

Seek medical advice before use. Guggulsterone may increase the risk of bleeding in people with **bleeding disorders**. Those with **hormone** sensitive conditions should avoid Guggulsterone; Guggulsterone may act like estrogen in the body and can make conditions like **breast cancer**, **ovarian cancer**, **endometriosis,** and **uterine fibroids** worse. If you have **hypothyroidism or hyperthyroidism, Guggulsterone** may interfere with treatment/medications. Animal studies raise vascular and liver-toxicity concerns; the mechanisms do not establish human benefit or safety. **Avoid self-treatment with heart or liver disease**.

Gymnema sylvestre

Gymnema sylvestre, from the *Apocynaceae* family where *periwinkles* and *toad plants* can be found, has been used in traditional Ayurvedic medicine. In Hindi it has been referred to as "*gurmar*" the sugar destroyer mainly since chewing on the leaves will temporarily leave you with the inability to taste sugar. Currently, it can be seen marketed as an anti-inflammatory, antidiabetic, and an anti-arthritic.

The active components of this plant are *gymnemic acids, gymnemasaponins,* and *gurmarin.* The fact that it can affect sugar mechanisms in the body could prove beneficial to diabetics.

- *Gymnemic acids* have been seen to stimulate insulin secretion in the pancreas; this has been seen to be accomplished via regeneration of *pancreatic islet cells; these* cells are involved in the activation of *beta cells (synthesize insulin)* and the inhibition of *alpha cells (elevates blood glucose levels).*
- *Gymnemic acids* and *gurmarin* have also shown a reduction in blood glucose levels. This is due to *Gymnemic acids* having similar structures to that of sugar molecules; this can lead them to "filling" sugar receptors and prevent the typical sugar activation in the body after consumption; this is the reason for not being able to taste sugars after chewing on the plant.
    - In the same manner, it is able to attach itself to the receptors in the intestines, reducing the sugar that is absorbed via the intestinal walls.

Gurmarin acts in a similar fashion, although it can also interfere with the taste buds' ability to tell bitter from sweet.

Gymnemasaponins, amongst other active components, have been shown to have anti-inflammatory effects that can also be seen to have anti-arthritic activity.

- The extract is able to reduce chronic swelling that has been correlated to bone destruction and reshaping as seen in arthritis. This may be due to the inhibition of *cytokines, GM-CSF (white blood cell growth factor, promoting rapid growth of immune cells)* and *interferons (involved in immune response and pain).* These are messengers in the immune system, responsible for inflammation and what causes discomfort and pain.

Benefits

- Fat loss (via sugar reduction)
- Antidiabetic
- Anti-arthritic
- Anti-inflammatory

Possible Side Effects

Not many side effects are known at this time, aside from low blood sugar, which can be seen as a benefit.

Dosage

Dosage is dependent on extract potency and brand; adhere to label directions, but most important of all, if using Gymnema to lower blood sugar levels, it should be done under the direct supervision of a medical professional.

WARNING

Please consult with a medical professional prior to administering Gymnema Sylvestre. There is concern that Gymnema Sylvestre may affect blood sugar in **diabetics**. Please consult with a doctor as medication may need to be modified.

Hawthorn Berry *(Crataegus pinnatifida)*

Hawthorn Berry, from the *Rosaceae* family where peaches, cherries, and almonds can be found, can also be known as *Chinese Hawthorn* and *Mountain Hawthorn*. Quick note, it has a history of carrying the name *Crataegus oxyacantha*, however, the name has been used for various species believed to be the same as *pinnatifida* and has since been rejected by the International Botanical Congress. Hawthorn Berry is native to Asia and has a long history of use in Traditional Chinese Medicine. The berry comes from the Hawthorn Shrub, and they're generally bright red. It is also believed that this is the shrub used to create Christ's crown of thorns, as the shrub is extremely spiny. Currently it can be seen marketed as a cardiovascular health aid, an anti-inflammatory that can also aid arthritic symptoms. Heart-disease evidence is limited or conflicting; hawthorn must not replace standard treatment, and product-specific interactions matter.

The main mechanism observed seems to come from a matrix of antioxidants found in the highest concentration within the leaves of the plant.

- Hawthorn berry extract has shown experimental inhibition of the absorption of triglycerides in the intestines, by inhibiting *pancreatic lipase (this is an enzyme responsible for the absorption and breakdown of triglycerides and fat molecules.)*
    - This can lead to lower *LDL* cholesterol, reducing plaque build-up, thereby enhancing vascular health.
- *Chlorogenic acid* is able to reduce *triglyceride* build up in the liver
- Another active molecule, hyperoside, has been found to be able to inhibit *NF-kB (where the stimuli for inflammation signals start)*, thereby reducing pro-inflammatory *cytokines (involved in cell signaling)* release. This can result in lower inflammation.
- *Epicatechin* in Hawthorn Berry seems to be potent at inhibiting *collagenase (enzymes that break down collagen structures)* and *gelatinase (enzymes that break down gelatin structures like collagen.)* This may aid those suffering from arthritis.

Pinnatifinoside seems to be a flavanol that is specific to this plant; however, there currently is no known human protein or gene interaction.

Proposed benefits (evidence varies)

- Cardiovascular health aid
- Anti-inflammatory
- Aid with arthritis

Possible Side Effects

- Nausea
- Upset stomach
- Sweating
- Headache
- Fatigue
- Dizziness

Dosage

- 23-35 mg a day for general health
- 100 mg a 3x a day for heart health

WARNING

Seek medical advice before use. Hawthorn Berry may affect people with **heart disease**. It may enhance or interfere with medication which could result in adverse effects or increased symptoms; please consult with a doctor prior to taking Hawthorn Berry.

HICA (*Alpha-hydroxy-isocaproic acid*)

HICA, also referred to as *leucic acid*, is a metabolite of **leucine**. It has been seen to contain *anabolic (the construction of molecules)* and *anti-catabolic (inhibits the breakdown of molecules)* properties. It is important to note that as of current, there is a lack of repetitive data, minimal testing has been done, and although it showed benefit, more research needs to be done to prove it to be conclusive. Currently, it can be seen marketed as a supplement that can increase lean muscle mass in athletes.

It does not seem to reduce *proteasome (breakdowns unneeded or damaged proteins)* or *Atrogin-1 (muscle-specific protein involved in muscle degradation for energy use)* activity. Its *anabolic* effect seems to be derived from its ability to sustain protein synthesis and *S6K1* in an otherwise *catabolic* environment.

- HICA can sustain *Ribosomal protein S6 kinase beta-1 (S6K1) activity; S6K1* is an enzyme whose *kinase activity (mechanism of transferring phosphate groups from high energy molecules)* leads to an increase in protein synthesis usually observed with physical exercise.
- HICA is also able to sustain protein synthesis, although not fully understood, it is believed to have similar if not the same mechanism as **Leucine** due to the fact that it is a metabolite of Leucine. In the same manner as in Leucine, protein synthesis will not happen unless the amino acids needed for protein synthesis are high enough, i.e., Leucine molecules. When these levels are met, the heterodimers are switched to "on" or active, starting a chain reaction in *mTORc1* that activates protein synthesis.
- HICA is believed to inhibit the enzyme *metalloproteinase*, which is also involved in the breakdown in muscle tissue; however, this is only speculated as it has been noted to reduce *DOMS (Delayed Onset Muscle Soreness). DOMS* usually appears within 24-48 hours after training; the reason behind *DOMS* is not yet conclusive, although it is believed that it might be linked to inflammation in response to muscle damage.
    - It seems by inhibiting *metalloproteinase* enzyme, this can potentially be where some of the main benefits of HICA come into play. By inhibiting *catabolism*, muscle mass can be preserved.

Benefits

- Increased lean muscle mass
- Reduce DOMS

Possible Side Effects

There are currently no known side effects at this time for lack of data or anecdotal reports.

Dosage

- 500 mg 3x a day

WARNING

Please consult with a medical professional prior to administering HICA. There are currently no known side effects at this time for lack of data or anecdotal reports. This does not mean that it's 100% safe; please monitor symptoms and report any unusual effects to your doctor. If HICA is anything like its counter-part leucine, **Diabetics** should proceed with caution as leucine has been seen to cause an acute release of insulin from the pancreas; please monitor your blood sugar levels.

HMB *(Beta-Hydroxy beta-methylbutyrate)*

HMB can be naturally found within the body as a down cascade metabolite of **Leucine** and is believed to be the main metabolite in Leucine that prevents muscle breakdown. It can also be found in foods such as avocados, asparagus, and cauliflower. It can also be found commonly referred to as *HMB-CA* or *HMB-FA* as these are variations of HMB. It can currently be seen marketed as a muscle enhancer with touted effects of increasing muscle size. Protein-turnover findings do not establish a universal advantage over *leucine*. Training studies are mixed, with possible differences by age, training status and regimen.

There appear to be two variations of HMB. *HMB-CA (a calcium salt-based form)* and *HMB-FA (a free acid form)*. When comparing *HMB-CA* to *HMB-FA*, *HMB-FA's* main win is that it is more bio-available, which allows for a quicker absorption rate in the body. Similar to **HICA**, one of the main benefits comes from its *anti-catabolic (inhibits the breakdown of molecules)* properties.

- *HMB* has also been seen to stimulate *mitogenesis (the induction of cell division)* in muscle tissue, thereby increasing muscle cell *proliferation (multiplying/reproduce.)*
    - o It is also able to increase protein synthesis through the *mTOR Pathway (indirectly affects protein function through phosphorylation, which can activate or deactivate protein enzymes).* Although *leucine acid* seems to be more effective at this, mainly due to the fact that there are various other factors involved in protein synthesis, in which Leucine's additional metabolites, *α-ketoisocaproic acid*, and *isovaleryl-CoA*, can both help mediate protein synthesis.

HMB has also been seen to reduce cell *apoptosis (programmed cell death)* of *myocytes (muscle cells)* while additionally increasing *MyoD (regulates muscle differentiation)* expression levels.

- It's been observed to accomplish this by increasing *BCL-2 (proteins that can cause or stop apoptosis)* as well as reducing *Bcl-2 associated X protein (BAX)* expression. *BAX* protein is involved and can activate *apoptosis* in a cell.
- *MyoD* is a protein that is generally seen after exercise or muscle damage. A sustained *MyoD* expression has been seen to also sustain the expression of genes involved in the maintenance of muscle.

Proposed benefits (evidence varies)

- Increased lean muscle
- Anti-catabolic
- May be anabolic

Possible Side Effects

There are currently no known side effects at this time for lack of data or anecdotal reports.

Dosage

- 1-3 grams a day

WARNING

Seek medical advice before use. There are currently no known side effects at this time for lack of data or anecdotal reports. This does not mean that it's 100% safe if HMB is anything like its counter-part *leucine*. **Diabetics** should proceed with caution as *leucine* has been seen to interact with insulin release.

Hoodia gordonii

Hoodia Gordonii may look like a cactus, but it's actually a leafless succulent from the *Apocynaceae* family, like Gymnema sylvestre (nearly all cacti are succulents, but not all succulents are cacti). Native to South Africa, it has a repulsive smell, comparable to that of rotten meat, and is pollinated by flies with surprisingly colorful plate-shaped flowers. The indigenous people of the area used this small desert plant for food, and as an emergency small water supply; they also believed it had medicinal properties. The plant has been used as an appetite suppressant. This led to the creation of the formula P57, to which Pfizer later released the rights due to a complex and difficult synthesizing process and toxic effects. Controlled human research has not established effective weight loss, and adverse cardiovascular and gastrointestinal effects have occurred.

It is believed the active component *P57* is what causes appetite suppression, although there is a debate about this, mainly on the basis that it is not able to cross the blood-brain barrier to theoretically have these effects. **It has been seen to have mild toxicity, especially to the liver, although it's inconclusive.**

- Theoretically speaking, if *P57* is able to reach the brain, the *hypothalamus,* in particular, it has been seen to increase intracellular **ATP** in *hypothalamic neurons*. The *hypothalamus* is directly responsible for appetite suppression. It is believed to have this effect as the *hypothalamus* may sense metabolic activity correlating it to nutrient/energy sufficient state, thereby producing appetite reduction appropriate to the ATP energy produced.

It is additionally believed that Hoodia Gordonii interacts with *beta-adrenergic receptors* as the toxic side effects of the extract mirror the effects of these receptors like decreased appetite, increased heart rate, and blood pressure to unsafe levels.

- *A beta-adrenergic agonist* can decrease food intake and suppress appetite; they also play a role in smooth muscle contraction and relaxation, which is the reason as to why heart rate and blood pressure can increase. It also interacts with neurotransmitters like *epinephrine* which might also be the reason for appetite suppression. The stimulation of *beta-adrenergic receptors* can also lead to cardiovascular symptoms.

Theoretically speaking, this may work, but *commercial* development did not establish an approved effective weight-loss medicine. However, it decided to stop production and return the rights to a supplement company; this should be of concern. There currently are safer alternatives for fat loss with fewer side effects.

Proposed benefits (evidence varies)

- May enhance fat loss
- Appetite suppressant

Possible Side Effects

- Increased blood pressure
- "Weird" skin sensations
- Vomiting
- Increased heart rate

Dosage

- No effective and safe weight-loss dose is established; traditional use and marketed doses are not clinical standards.

WARNING

Seek medical advice before use. Avoid self-treatment with **heart disease**, **hypertension**, **diabetes, high cholesterol** or liver disease. Human research reports adverse cardiovascular and gastrointestinal effects; animal and laboratory toxicity concerns add uncertainty. Proposed appetite suppression does not establish a safe weight-loss **treatment**.

Hordenine *(N-dimethyltyramine)*

Hordenine is an alkaloid that can be found in *bitter orange (a supplement associated with similar effects)* extract, amongst other plants. Its name is derived from barley *(Hordeum vulgare)* as it can be found present in germinated barley which has been implicated to the mood-elevating effects in beer. Mainly because hordenine is a strong *Dopamine D2 receptor* agonist due to having an almost identical *ligand* as *dopamine.* It has an *adrenergic* like structure which may be the reason as to how it's currently marketed, a nutritional supplement claiming to be a stimulant and a weight loss product. Human efficacy and safety evidence is inadequate. FDA has identified hordenine in recalled products as outside the statutory dietary-ingredient definition.

A proposed *lipolytic* action would release fatty acids from stored triglycerides, not release whole fat cells. This does not establish improved body composition in people.

- In performance, it seems it is able to enhance *adrenaline-induced* contractions by inhibiting *noradrenaline reuptake.* Simply put it this way, when *noradrenaline reuptake* is inhibited, this means that there is more *noradrenaline (neurotransmitter)* sitting around cells because the neurotransmitters are not reabsorbed.
    - Additionally, it has been seen as a *norepinephrine (a hormone that excites neurons)* stimulant as it is believed to release *norepinephrine* from its stores; in short, it means it could lead to improved muscle contractions *(inotropic).*
- Since it seems to act as an *adrenergic (works via adrenaline)* compound, effects similar to that of *adrenaline* may be observed.
    - Meaning fat loss may be achieved through the same functions as *adrenaline.*
- This would, in turn, result in a release of *glucose* from energy stores and increased blood flow to skeletal muscles while reducing blood flow to the gastrointestinal system.
    - In simpler terms, reduced blood flow to the gastrointestinal system theoretically means decreased hunger and slower food digestion, along with improved blood flow to muscles.

Now where fat loss comes into play would be *glucose* release. Putting all this together means the lack of food consumption would lead to a lack of carbohydrates needed for glucose, which, in turn, would result in *gluconeogenesis (GNG). GNG* is the creation of *glucose* from non-carbohydrate compounds. One of these compounds is *lipids* like *triglycerides,* which in part make up body fat.

Proposed benefits (evidence varies)

- May increase power output
- May aid fat loss
- May enhance body composition

Possible Side Effects

- Increase heart rate
- Increased blood pressure
- Shortness of breath
- Sweating

Dosage

No established safe and effective human performance dose exists; a label recommendation does not resolve the evidence or regulatory concerns.

WARNING

Seek medical advice before use, especially with **heart disease**, **hypertension or diabetes. Stimulant effects can affect heart rate, blood pressure** and glucose. Reports of drug-test interference require confirmation; h**ordenine** is not an opiate. Safety of an exercise regimen remains unestablished.

Horny Goat Weed *(Epimedium)*

Horny Goat Weed generally refers to the *genus* of hardy evergreen flowering plants in the *berberidaceae* family rather than one specific species. It is native to Asia and some parts of the Mediterranean. It has been used in traditional Chinese Medicine as an erectile aid and aphrodisiac. Currently, it can be seen marketed as a testosterone booster, nootropic, and heart health supplement. Interestingly enough, it seems to be selective based on gender-specific hormones after ingestion, for example, it not only increases testosterone in men but estrogen in women too.

The bioactive component is *Icariin* which is extremely unstable in the body and generally needs to be combined with a *lactase (lactase phlorizin hydrolase)* inhibitor to promote bioavailability and absorption in the intestines.

- *Icariin* has been seen to aid with *osteoporosis* prevention by reducing bone mineral density loss. It accomplishes this via *osteogenic differentiation (the creation of bone cells from mesenchymal stem cells)* and inhibiting *osteoclast differentiation (the creation of specialized cells that degrade bone matrix) and bone resorption (the breakdown of bone tissue and release of minerals into the body).*
- It has also been seen to inhibit the *Acetylcholinesterase enzyme (AChE),* which has been linked to the breakdown of the neurotransmitter *acetylcholine, which would* promote cognitive function.
    - This may also aid in the prevention of neurodegeneration. It has also been seen to increase *neuronal nitric oxide synthase (nNOS), improving blood flow* to the nervous system, which could, in turn, promote nerve growth.
- It can also increase *endothelial nitric oxide synthase (eNOS),* enhancing blood flow in the body.

Icariin seems to be able to, in a dose-dependent manner, stimulate *Leydig cells (responsible for testosterone production)* without affecting *Luteinizing Hormone (LH).* This is believed to be accomplished as *Icariin* has testosterone mimetic (similar/imitate) properties which can promote testosterone secretion via increasing *Cyclic adenosine monophosphate (cAMP)* levels. *cAMP* plays a role in transferring the effect of hormones and the activation of certain enzymes involved in cellular processes *(protein kinases).*

Benefits

- Testosterone booster
- Vasodilator
- Erectile aid
- Enhance osteoporosis treatment
- Nootropic

Possible Side Effects

- Dizziness
- Nosebleeds
- Dry mouth
- Vomiting
- Spasms

Dosage

Dosage varies by extract concentration, at 40% extract 80 mg per 2.2 lbs/ 1 kg; adhere to label instructions

WARNING

Please consult with a medical professional prior to administering Horny Goat Weed. There is concern that Horny Goat Weed may increase the risk of bleeding in those diagnosed with **bleeding disorders**. If you have a hormone sensitive condition, such as **breast, uterine** or **ovarian cancer**, Horny Goat Weed can worsen these conditions. If you have **low blood pressure,** Horny Goat Weed can make blood pressure too low. There currently is a lack of data and more research needs to be done to be conclusive.

Hyaluronic Acid

Hyaluronic acid, also called hyaluronan, is a *glycosaminoglycan (GAG); GAGs* are vastly important in the make-up of *extracellular matrix,* like that found in joint cartilage. Hyaluronic Acid is found in connective, neural, and *epithelial (tissue that lines blood vessels and organs)* tissues within the body. It also plays a part as a lubricating fluid between joints and is one of the main components in cartilage. Currently, it can be seen marketed as a treatment for osteoarthritis, and like many cartilage aid products, it can also be used to aid skin aesthetic appearance.

It is important to note that *magnesium* plays a role as a rate-limiting nutrient for Hyaluronic Acid, meaning if there is not enough *magnesium* in the body, Hyaluronic Acid production may suffer. Be sure to supplement *magnesium* properly in your diet along with Hyaluronic acid to achieve optimal effects.

- Many of Hyaluronic Acid's benefits come from its binding affinity (not in the sense of your typical binding) to *Toll-like Receptor 4 (TLR4). TLR4* is responsible for inflammatory responses via intracellular signaling of *NF-kB (responsible for cytokine production).* Hyaluronic Acid is then able to modulate the inflammatory process, mainly as it thickens the jelly shield around the cell regulating access through the receptor via a barrier, thereby reducing the inflammatory response.

Another way Hyaluronic Acid can provide benefits is through the same function as **glucosamine**. Hyaluronic Acid is made up of *N-acetyl glucosamine* and *D-glucuronic Acid (makes up part of chondroitin).*

- **Chondroitin** and glucosamine have been known to promote *collagen synthesis in* ligaments, *chondrocytes (cartilage cells)* and *tenocytes (tendon cells).*

There is a good amount of data for positive benefits for those suffering from osteoarthritis, although it's not as quick-acting as steroids, its effect builds at a slower pace but believed to have longer-lasting relief.

- Hyaluronic acid plays an important role in skin moisture, as it is able to retain water molecules and protection. This means it can promote scar-free tissue repair, as well as possibly reduce aging signs.

Hyaluronic acid has a half-life of 3 to 5 minutes on the skin. Topical application may provide benefits, as mentioned above, supplementing *magnesium* will promote the synthesis of Hyaluronic Acid.

Benefits

- Aid in osteoarthritis treatment
- Improve skin appearance

Possible Side Effects

- Photosensitivity after topical application

Dosages

Dosages vary in application methods, injection or oral administration.

- The research suggested oral administration 200 mg a day

WARNING

Please consult with a medical professional prior to administering Hyaluronic Acid. Currently, there are no reported side effects through oral administration. Although injections should solely be done by a professional medical provider, as these can result in possible complications and can lead to muscle pain and stiffness or a burning sensation in the location where the shot was administered.

Inositol *(myo-Inositol)*

Inositol is a small sugar alcohol and is similar in structure to *glucose.* It is synthesized in the pancreas from *glucose* and can be found throughout the body as it does play a role in cellular messaging, with the brain having the highest levels. It is additionally involved in the response of hormones and growth factors which may be the reason as to its current aid with *Polycystic Ovary Syndrome (PCOS).* It has also been seen to be a possible *anxiolytic (anti-anxiety)* as well as a possible aid in insulin resistance.

PCOS seems to be an *endocrine (hormone)* disorder. In the ovaries, Inositol is involved in *glucose* uptake and *follicle stimulating hormone (FSH)* signaling, while *D-chiro-inositol (synthesized from inositol)* is involved in *androgen* synthesis via insulin. An imbalance of inositol can affect hormone signaling, specifically *FSH,* which has been observed in women with *PCOS. FSH* is involved in the growth of ovarian follicles and *estradiol (major female sex hormone)* production. This imbalance can negatively affect the metabolism of both *follicular* cells and *oocytes* as *oocytes (involved in fertility and egg development)* have a high rate of *glucose* consumption.

- Abnormal *FSH* levels can prevent ovarian follicles from maturing and releasing the egg cells creating the cysts seen in *PCOS* due to the accumulation of immature *ovarian follicles.*

Inositol and *D-chiro-inositol* can also affect insulin response as they are part of *inositol phosphoglycans (an insulin messenger).* It is believed that *inositol phosphoglycans* directly control secondary actions of insulin which means that insulin impairment may be due to *inositol phosphoglycans* imbalance.

- This means that Inositol imbalance can additionally lead to insulin signaling dysregulation, which means it would not only affect *PCOS* but also cause insulin resistance, especially in those with *PCOS.*

For women suffering from *PCOS, Inositol* may provide benefits. Supplementation of Inositol can very well improve ovarian function and metabolism as well as hormonal regulation.

In some neurological brain conditions, like Alzheimer's, Inositol levels have been seen to be negatively lowered. Inositol supplementation may help as inositol can be found in *glial (support and protect neurons)* cells in the brain where it acts as an *osmolyte (maintains proper cell fluid and volume in cells).*

Benefits

- Aid with *PCOS*
- Antidiabetic
- Anxiolytic

Possible Side Effects

- Dizziness
- Tiredness
- Headache
- Stomach discomfort
- Nausea

Dosages

- 12-18g a day for anxiolytic effects
- 2-4g for PCOS specifically *myo-Inositol* and *D-chiro-inositol*, once a day before breakfast

WARNING

Please consult with a medical professional prior to administering Inositol. There is concern that Inositol can affect **diabetics** as it can affect insulin response; medication may need to be modified. For those with bipolar disorder and depression, Inositol may worsen conditions as it can affect neurotransmission in the brain, possibly leading to a manic episode.

Ipriflavone *(7-isopropoxyisoflavone)*

Ipriflavone is a synthetic version of an *isoflavone; isoflavones are generally regarded as naturally occurring* phytoestrogens generally derived from the *Fabaceae* family, commonly referred to as the bean family. Although it has been commonly referred to as a *phytoestrogen*, its mechanisms and interactions within the body seem to differ, showing no signs of interaction with *estrogen* receptors or *estrogenic* effects. Currently it can be seen marketed as an anabolic supplement touted to build muscle (anecdotal data state this ineffective); however, it can also be seen marketed as an aid for postmenopausal *osteoporosis.*

Ipriflavone seems to relay its effect mainly in bone formation as a reduction in *hydroxyproline (amino acid involved in fibrillar collagen-like that found in bone)* has been noticed with use. However, other studies have shown the opposite effect with the addition (side effect) of many women developing *lymphocytopenia (condition of low levels of lymphocytes in the blood).*

- A substantial randomized osteoporosis trial did not establish benefit and reported lymphocytopenia. Seek established osteoporosis assessment and treatment rather than substituting another unproven supplement on the basis of a bone-metabolism mechanism.

Its main mechanism, however, is believed to inhibit *parathyroid hormone* and *PGE2* mediated bone resorption.

- *Parathyroid hormone* influences bone remodeling via regulation of calcium levels in the blood; this directly affects bone resorption (the breakdown of bones), since bones contain large amounts of calcium.
- There currently is debate in regard to whether or not *Prostaglandin E2 (PGE2)* causes bone formation or bone resorption, so once again its mechanism behind bone sustainment through *PGE2* is obscure. Currently, it is also believed that *PGE2* has a stimulatory effect in *bone marrow-derived stromal cells osteogenesis (differentiation of bone cells from stromal cells).* However, it is also believed that *PGE2* shows preference in facilitating *adipogenesis (differentiation of fat cells)* over *osteogenesis*, possibly by reducing *osteocalcin (plays a role in pro-bone building homeostasis); this* can negatively affect bone growth and sustainment.

In a nutshell, it seems *PGE2* can positively affect bone formation under the right circumstances; however, under other circumstances, such as a hormonal disbalance or *dexamethasone (anti-inflammatory corticosteroids drug)* use, it can favor the creation of *corticosteroids.*

Proposed benefits (evidence varies)

- May aid osteoporosis
- May increase bone density

Possible Side Effects

- Gastrointestinal discomfort
- Diarrhea
- Dizziness

Dosages

- A major trial used 200 mg three times daily, but found no osteoporosis benefit and blood-cell risks; this is not a recommended self-treatment regimen.

WARNING

Seek medical advice before use. Ipriflavone can severely affect those with **compromised immune systems** as it has been known to lower White blood count.

Iron

Iron is a chemical element found on the Periodic Table sitting at atomic number 26. It is in the Transition Metal group where **Manganese** and **Chromium** are found, and it is a malleable silvery-gray metal. Iron is the most abundantly found metal element on earth, making up much of the earth's inner and outer core. In our bodies, it has a vital role; it aids in forming *hemoglobin (a protein that transports oxygen in blood cells)* and *myoglobin (carries oxygen in muscle tissues).*

Since iron is vital to cell function, a deficiency can lead to fatigue/mental fogginess; this may be due to *hemoglobin (involved in oxygen transportation)* levels. It can also compromise immune system health which can lead to you feeling cold constantly as seen in *anemic (state in which there's not enough healthy blood cells or hemoglobin to carry oxygen)* states. Calcium or dairy products, as well as phytates (bind to minerals), which are commonly found in the majority of plant foods, can affect iron absorption. It is also important to note that there are 2 different types of iron, *heme iron*, and *non-heme iron*

- Heme iron is iron bound within heme, found in animal tissues. It is generally absorbed more efficiently than non-heme iron; absorption varies with iron status and the meal. Heme does not mean non-synthetic.
  - Consuming Vitamin C alongside non-*heme iron* in a meal can increase absorption in the body.
- Non-heme iron occurs in plant foods, fortified foods, animal foods and most supplements. It is not simply synthetic iron. Its absorption is more sensitive to food components a*nd iron* status than *heme iron* absorption.
  - Usually, these iron pills will say something along the lines of "ferrous sulfate" or "ferrous fumarate."

Generally, some meal components, such as polyphenols, can reduce the absorption of iron, as many antioxidants have *chelating properties; chelation* is the ability to bind to heavy metals in the body making them easier for the body to remove. This does not mean that all antioxidant-rich foods prevent iron absorption.

Benefits (Benefits seen are for those in deficient states)

- Increased
- Increased mental alertness

Possible Side Effects

- Upset stomach
- Vomiting
- Nausea
- Diarrhea
- Constipation

Dosages

- Adult RDA: 8 mg/day for men and adults 51+. Training alone is not a reason to raise the dose without assessment.
- 18 mg/day for women ages 19–50; pregnancy: 27 mg/day.
- 15 mg/day for girls ages 14–18; treatment of deficiency differs.

WARNING

Seek medical advice before use. High iron intake is one of the most common causes of fatal poisoning amongst children. Iron poisoning can cause **stomach** and **intestinal distress**, **liver failure,** and extremely **low blood pressure**. It may also aggravate conditions such as **intestinal ulcers** and **inflammation**. If you are diagnosed with a **hemoglobin disease,** iron supplementation can cause an iron overload.

Jojoba Oil *(Simmondsia chinensis)*

Jojoba is a plant that can be mainly found in the west southern part of the United States like Arizona and Southern California, as well as Northern Mexico. It belongs to Simmondsi*aceae in* the order *Caryophyllales*. The supplement is mainly sold as a light-colored oil that's extracted from the jojoba seed. It has a long history of use in traditional Native American medicine generally to treat sores and wounds. It is also used in cosmetic and skincare products as the oil has skin health benefits. It is a liquid wax of long-chain fatty acid and fatty alcohol esters, used topically rather than as an edible vegetable-oil substitute; however, *the appetite suppressing effect that it can be seen marketed for is not due to the oil* itself but Jojoba *simmondsin*. Simmondsin appetite findings are mainly experimental; they do not establish a safe ora*l supplement or weight-loss regimen.*

Simmondsin (a glycoside found in the seeds) seems to suppress appetite by indirectly interacting with the *Vagus Nerve* via *cholecystokinin (CCK) receptors*. The *Vagus Nerve* is responsible for gastrointestinal muscle contraction and relaxation in the intestines. *CCK* is a hormone that is released in the presence of fatty acids and broken-down protein in the gut; CCK is then able to activate *CCK receptors* thereby activating the *Vagus* terminals through an *intracellular* calcium increase; this has the ability to suppress hunger.

Jojoba oil or liquid wax is able to enhance skin recovery and aesthetics by enhancing *fibroblasts* and *keratinocytes* capabilities through the *PI3K/Akt/mTOR pathway; this* pathway regulates a cell's cycle.

- *Fibroblast* is a cell that is able to synthesize collagen as well as being critical to wound healing by responding to tissue injury.
- *Keratinocytes* are responsible for protecting the skin by forming a barrier to protect against environmental damage such as bacteria, UV radiation, water loss, etc. If anything penetrates the skin, *keratinocytes* can activate *chemokines (signaling proteins)* to attract *leukocytes (white blood cells)* to the area of invasion.

This can result in enhanced skin aesthetics, like reduced signs of aging, lesions, and scar tissue formation on the skin through topical application.

Proposed benefits (evidence varies)

- Enhance skin appearance
- Appetite suppressant
- Aid fat loss

Possible Side Effects

- May cause rash at the site of topical application

Dosages

Topical products should be used as labeled. Jojoba is not a routine edible oil; no safe effective oral weight-loss dose is established.

WARNING

Seek medical advice before use. Jojoba Oil may affect those with **heart conditions** as it contains *erucic acid,* an omega-9 fatty acid. *Erucic acid* was seen to have toxic effects on the heart at high doses in animals. Although no link between heart problems and *erucic acid* have been found in humans, along with other oils/products that contain *erucic acid* to produce any sort of harm. It is advised to proceed with caution as the full side effects and benefits of Jojoba Oil are not known.

Kelp (*Ascophyllum nodosum*)

Kelp is a general term referring to various strands of seaweed, each having their own individualized health effects/benefits. *Ascophyllum nodosum* is the main species seen in supplements; this strand is a cold-water loving seaweed native to the Northern Atlantic Ocean. It is a tough leathery seaweed with a slow growth rate reaching up to 2 meters. It can be seen currently marketed as an immune system booster as well as having the capability to inhibit carbohydrate absorption.

The active components in this strand of kelp are phlorotannins *(a marine-exclusive polyphenol)* and *ascophyllan.*

- *Phlorotannins* seem to have the ability to inhibit starch and lipid uptake via *α-glucosidase* and *α-amylase* inhibition. *α-glucosidase* and *α-amylase* break down starches and saccharides (sugar molecules) into glucose, which can lead to lower glucose levels. This results in lower insulin levels through enhanced insulin sensitivity.
- *Phlorotannins* have also been seen to reduce overall *Reactive Oxygen Species (ROS)* and *lipid peroxidation* while also increasing **glutathione** levels and augmenting their activity; this has shown promise as an effective antioxidant significantly reducing oxidative damage.
    - *Phlorotannins* have additionally been seen to reduce inflammatory responses that can be synergistic with its antioxidant properties. It can downregulate inflammatory gene expression for the *cytokines (inflammation messengers) Interleukin-8, Tumor Necrosis Factor-α* just to name a few. Meaning this can reduce the overall creation of pro-inflammatory messengers prior to the event taking place.
- *Ascophyllan* has been seen to stimulate *phagocytosis (a phagocyte cell that consumes bacteria or harmful substances)* while also having *chelating* activity. *Chelation* is the bonding of toxic metal ions to a molecule making them easier for the body to remove, thereby preventing damage to cells

Although the effect is still debated, it is believed that supplementing this with a meal may reduce the amount of food consumed, which may lead to overall weight loss; the reasons as to why have not yet been verified.

Proposed benefits (evidence varies)

- Enhance immune system
- Aid in weight loss
- Possible antidiabetic
- Anti-inflammatory
- Antioxidant
- Chelator

Possible Side Effects

For lack of data or anecdotal reports, side effects have not been reported.

Dosage

- 4.6 grams a day with food

WARNING

Seek medical advice before use. Kelp may affect **diabetics** as it interacts with insulin sensitivity and glucose uptake. Those with **thyroid** and **kidney dysfunction**, Kelp is high in iodine and potassium, which can stress the kidneys and affect Thyroid function due to the high amounts of iodine. Large amounts of Kelp can contain arsenic. Iodine content varies widely; excess can cause thyroid dysfunction, so weight of kelp is not a reliable iodine dose.

Kola Nut *(Cola nitida)*

Kola Nut, not to be confused with **Gotu Kola**, is a nut from the Kola Tree in the *Malvaceae* family where the cocoa tree, *Theobroma*, can be found. It is native to Tropical Africa; it is an evergreen tree that averages around 20 meters in height. Although most people are unfamiliar with the nut or the tree, most may recognize the taste as it is used to flavor Cola. Kola nuts tend to have a high caffeine content and have been generally marketed as a fat-loss supplement and as a general stimulant like caffeine.

Its fat loss effects may come from inhibition of *α-glucosidase (found in the small intestine)* and *α-amylase (can be found in saliva)* inhibition. *α-glucosidase* and *α-amylase* break down starches and saccharides (sugar molecules) for the absorption and creation of glucose.

- Inhibition of these two enzymes can allow various sugars to pass through the body, reducing overall absorption which can lead to lower blood glucose levels and possibly enhance insulin sensitivity. In the same fashion it may also reduce overall fat storage indirectly due to the reduced amount of sugars taken up by the body, meaning fat storage due to excessive glucose in the body will be reduced.
- Kola Nut extract has also been seen to promote antioxidant function, mainly through its *phenolic* compounds, by raising *glutathione catalase* and *Superoxide Dismutase* levels. These antioxidants are *endogenous (from within the body)* and play important roles in protecting cells from oxidative damage.

The use of the nut and its extracts have been seen to negatively affect male fertility. The cause of this is not fully understood; however, it is believed that it is able to down-regulate male reproductive hormones, *Luteinizing Hormone,* and *testosterone,* that affect *spermatogenesis (creation of sperm).* It has been additionally seen to be a powerful *diuretic* removing potassium (kaliuretic) and sodium (natriuretic) from the body through urine. This can be seen as an undesired effect for some especially, those with low blood pressure as sodium is involved in blood pressure and potassium is involved in blood circulation.

Benefits

- Antidiabetic
- Antioxidant
- Fat loss aid

Possible Side Effects

- Nausea
- Insomnia
- Gastrointestinal discomfort
- Irritation
- Vomiting
- Anxiety
- Headache
- Agitation
- Increased heart rate
- Restlessness
- Dizziness
- Ringing in ear

Dosages

No set dosage is known due to lack of data and research, adhere to label instructions

WARNING

Please consult with a medical professional prior to administering Kola Nut. There is concern that Kola Nut may affect those with **bleeding disorders** as Kola Nut has been known to affect clotting. Those with **high blood pressure**, Kola Nut can affect blood pressure due to the caffeine content and its diuretic effects. For those with heart conditions, Kola Nut can affect heart function due to its effect on electrolytes. Kola Nut can affect **Diabetics** due to its interaction with glucose; insulin medication may need to be modified.

Krill Oil *(Euphausia superba)*

Like other fish oils, Krill Oil has been seen to be cardioprotective and promote cognitive health; however, Krill Oil is more easily absorbed by the body than regular **Fish Oil**. Krill Oil has *omega-3* fatty acids, *eicosapentaenoic acid (EPA)* and *docosahexaenoic acid (DHA),* mainly in the form of *phospholipids* which results in a higher bioavailability compared to Fish Oil, but absorption differences do not establish superiority to Fish Oil for every clinical outcome. Those that do not regularly consume fish in their diet can see an increase in cognitive function from DHA; however, Krill Oil *DHA* seems to have increased neuroprotective effects from its *phospholipid* make up compared to regular Fish Oil *DHA*. *EPA* can increase blood flow to the brain thereby enhancing *DHA's* effect. Through this synergy, it has been seen to aid those with diagnosed cognitive problems. Generally speaking, *those that consume enough fish might not see the benefit of this supplement*.

Enhanced *EPA* concentrations in *Neutrophils* may be the reason for its anti-inflammatory effects. *EPA* seems to *attenuate (slow/reduce) leukotriene B4 (inflammatory signaler)* production in *neutrophils (a type of white blood cell)*. It has also been seen to reduce inflammation through the suppression of gene transcription for *phosphoinositide 3-kinase (PI3K)* activation. *PIK3* controls intra-cellular signaling and modulation of lipids like **arachidonic acid (AA)**. Modulating the production of *eicosanoids* can either inhibit or cause inflammation. It can also reduce inflammation through modifying the production of eicosanoids; these are messengers that are released after **arachidonic acid** or polyunsaturated fatty acids are broken down; they either inhibit or cause inflammation.

- Through this process, it is thought to be able to reduce plaque build-up in the arteries, increasing cardiovascular health and possibly reduce joint inflammation especially those with osteoarthritis.

Krill Oil and Fish Oil have been able to reduce PMS symptoms; it is believed to work through *G_s alpha subunit (a G Protein that activates and is involved in neurotransmission like dopamine & serotonin)* modifying, allowing *G proteins* to more effectively activate receptors on the cell. Animal gene-expression comparisons do not establish greater human benefit or equivalent doses; results depend on the *preparation* and endpoint.

Benefits

- Arthritis Aid/joint health
- Aid cardiovascular health
- Anti-inflammatory
- Enhanced cognitive function
- Neuro-antioxidant

Possible Side Effects

- Low blood pressure
- Fish "burps"
- Nausea
- Heartburn

Dosages

- Compare actual EPA+DHA content, not oil weight alone. No universal two-thirds conversion from Fish Oil is established.

WARNING

Seek medical advice before use. This brand of Krill Oil may affect those with **Seafood/Shellfish allergies**, as Krill Oil may cause an allergic reaction. Those diagnosed with **bleeding problems**, Krill Oil can increase the risk of bleeding. Krill Oil may make blood pressure too low for people with **low blood pressure.**

L-Arginine

L-Arginine is a semi-essential amino acid; it has been mainly used for maintaining nitric oxide levels and blood flow in athletes. L-Arginine has poor bioavailability; supplementation may not increase arginine levels as much as **L-Citrulline**, as L-Citrulline is converted into arginine in the body.

Ultimately, arginine supplementation can acutely increase *growth-hormone (GH)* during times of rest.

- Its *GH* effects have mainly been noted through increased calcium influx to *pituitary* cells which are believed to increase the expression of *GH in the body; this* effect seems to be dose-dependent.

Arginine is extremely versatile in the body as it has the ability to convert into **creatine** or **agmatine** and is also involved in the *urea cycle* and *nitric oxide cycle.*

- In the *Urea Cycle,* arginase converts arginine to ornithine and releases urea. Ornithine combines with carbamoyl phosphate to form citrulline; ammonia contributes nitrogen through carbamoyl phosphate, not as a cofactor attached directly to ornithine. The cycle removes nitrogen as less-toxic urea. Its normal function does not establish improved athletic performance from added arginine.
    - The cycle then turns citrulline into argininosuccinate; argininosuccinate can then be converted back to arginine, which would result in free arginine in the body that could either be used for the *Urea Cycle* or the *Nitric Oxide Cycle*
- In the *Nitric Oxide Cycle*, similar to above, arginine is converted into citrulline through *Nitric Oxide Synthase (NOS)* enzymes. Citrulline can be converted back to arginine via *argininosuccinate.*
    - There are 3 types of NOS: *endothelial NOS (eNOS)* blood vasodilator, *neuronal NOS (nNOS)* nervous system/neurotransmitter development *and inducible (iNOS)* involved in immune response such as inflammation.

This mechanism promotes nNOS; this can result in enhanced cognitive function, as it stimulates synaptic neurons, enhancing transmissions which can result in better memory formation with the addition of proper motor function in the muscles. *eNOS* is able to relax smooth muscles, promoting vasodilation, resulting in enhanced blood flow throughout the body to include the brain and muscle tissue.

Proposed benefits (evidence varies)

- Cognitive enhancement
- Cardiovascular health aid
- Enhance *GH* expression
- May increase endurance
- Vasodilator
- Pro-inflammatory

Possible Side Effects

- Abdominal pain
- Low blood pressure
- Gout
- Diarrhea
- Inflammation

Dosages

- 3-6g a day

WARNING

Seek medical advice before use. L-Arginine may affect people with inflammatory conditions such as **asthma** and **allergies,** as L-Arginine can promote inflammation responses. Those diagnosed with **low blood pressure,** or heart failure, L-Arginine can lower your blood pressure to unsafe levels. **Guanidioacetate methyltransferase deficiency** and **herpes** can be aggravated after the supplementation of L-Arginine. Do not self-supplement after a heart attack: the VINTAGE MI trial stopped early after excess deaths in the arginine group.

L-Arginine Ethyl Ester (LAE)

Arginine is a semi-essential amino acid that has been mainly used for maintaining nitric oxide levels and vasodilation in athletes. Arginine has poor bioavailability and is poorly absorbed as it is quickly broken down in the intestines, leading to either a minor effect in the body or needing higher Arginine doses. It is believed that the attachment of an *ester (an organic compound)* to arginine could enhance absorption rates by maintaining arginine intact in the intestines thereby making Arginine effects more pronounced.

Arginine and Arginine Ethyl Ester share an arginine component, but equivalent efficacy or improved absorption from esterification is not established. They are both involved in the *Urea (removal of ammonia)* and *Nitric Oxide Cycle; the addition* of the ester is the only change.

- In the *Urea Cycle,* arginase converts arginine to ornithine and releases urea. Ornithine combines with carbamoyl phosphate to form citrulline; ammonia contributes nitrogen through carbamoyl phosphate, not as a cofactor attached directly to ornithine. The cycle removes nitrogen as less-toxic urea. Its normal function does not establish improved athletic performance from added arginine.
    - The cycle then turns citrulline into argininosuccinate; argininosuccinate can then be converted back to arginine, which would result in free arginine in the body that could either be used for the *Urea Cycle* or the *Nitric Oxide Cycle*
- In the *Nitric Oxide Cycle,* similar to above, arginine is converted into citrulline through *Nitric Oxide Synthase (NOS)* enzymes. Citrulline can then be converted back to arginine via *argininosuccinate.*
    - There are 3 types of NOS: *endothelial NOS (eNOS)* blood vasodilator, *neuronal NOS (nNOS)* nervous system/neurotransmitter development *and inducible (iNOS)* involved in immune response such as inflammation.

This mechanism promotes nNOS; this can result in enhanced cognitive function, as it stimulates synaptic neurons, enhancing transmissions which can result in better memory formation with the addition of proper motor function in the muscles. *eNOS* is able to relax smooth muscles and promote vasodilation, resulting in enhanced blood flow throughout the body to include the brain and muscle tissue.

<u>Proposed benefits (evidence varies)</u>

- Promote Muscle growth
- Aid Erectile Disfunction
- Vasodilator
- Promote muscle repair
- Antioxidant
- Improve strength

<u>Side Effects</u>

- Low blood pressure
- Abdominal pain
- Abdominal Pain
- Bloating
- Diarrhea

<u>Dosages</u>

- Recommended daily dosage is 1-4 grams

WARNING

Seek medical advice before use. Arginine Ethyl Ester may affect people with **Kidney** disease as Arginine can increase potassium levels. Those with **cardiovascular** problems such as **strokes** and **blood pressure problems, Arginine** can lower blood pressure to unsafe levels. Those with **asthma** or **allergies**, Arginine could increase inflammation in the airways due to an allergic response or asthma attack. Post-heart-attack arginine safety concerns also apply; esterification does not establish protection from these risks.

L-Carnitine

There are various forms of carnitine, L-Carnitine being the most basic. It's synthesized in the body through **lysine** and **methionine**. It can stimulate *Nitric Oxide Synthesis (NOS)*, as well as act as an antioxidant. The body can convert L-carnitine and **Acetyl-L-Carnitine (ALCAR)** to the necessary derivative dependent on need. It is important to note that if a Vitamin C deficiency exists, the synthesis of carnitine will be impaired.

Carnitine can increase glucose and creatine levels in the brain which can reduce cognitive decline, as well as increase serotonin levels, which can be *anxiolytic (anti-anxiety)* and elevate mood.

- Carnitine is believed to be implicated as a possible antioxidant of *Nitric Oxide* as it has been seen to be able to stimulate *Glutathione* expression as increased glutathione levels in the blood can reduce *nitric oxide* breakdown. Carnitine is also believed to promote *Nitric Oxide* expression at a transcriptional level possibly relating to *endothelial Nitric Oxide Synthase (involved in vasodilation).*

Carnitine is linked to enhanced performance. It is able to accomplish this by reducing *mitochondria* decay in muscle. Since carnitine aids in mitochondrial transport of *fatty acids (fat molecules)* for fuel, *adipose (fat)* tissue can then be used for **ATP** in the mitochondria.

- Generally seen, L-Carnitine, aids with *nitric oxide synthesis,* although GPLC is more efficient. Nitric *oxide* can be enhanced following a carb-rich meal; this aids the transport of *carnitine* into muscle tissues.
- It has also been suggested that Carnitine can reduce lactate build-up in muscle tissue due to rigorous exercise; however, studies to date have shown no effect of lactate levels in muscle tissue nor any alteration.

Not only is this beneficial through exercise, but it also has the potential to tremendously aid recovery. Blood flow is maintained after exercise; through a proper nutritional program and post-workout nutrition, blood is able to transport nutrients more efficiently, meaning efficient protein synthesis, which means increased muscle mass which can result in a reduction of feelings of soreness.

Benefits

- May promote fat loss
- Vasodilator
- Increase endurance
- Nootropic (cognitive)
- Aid recovery

Possible Side Effects

- Fluctuating blood pressure
- Headache
- Nausea
- Vomiting
- Diarrhea
- Stomach cramps

Dosage

- The recommended dosage is 500-2,000 mg a day

WARNING

Please consult with a medical professional prior to administering L-Carnitine. There is concern that L-Carnitine can affect those with **thyroid problems** as it has been seen to interact with Thyroid hormones. L-Carnitine has also been linked to an increase in seizures in those already prone to **seizures**. If you are under **blood thinners** or **blood pressure** medication, L-Carnitine has been known to affect clotting while also reducing blood pressure.

L-Cysteine

First and foremost, an issue needs to be addressed in regard to L-Cysteine and **NAC** benefits. These two supplements, although similar, are **NOT** interchangeable. The majority of benefits that you will read about in other sources will put L-Cysteine and NAC as the same. *First Pass Metabolism (drug breakdown by the body)* of L-Cysteine minimizes distribution in the body. The *acetylated* cysteine diminishes this breakdown effect, increases absorption in the intestines, and allows it to cross the blood-brain barrier. There are benefits to L-Cysteine supplementation, but they are minimal at best, and cannot replicate NAC benefits. With that being said, this page will focus on the benefits and effects of L-Cysteine.

L-cysteine contains a *thiol (high energy molecule)* that allows it to help various biological reactions in the human body. L-cysteine has been implicated, not as effectively as NAC; however, at supporting the synthesis of **glutathione**. Maintaining proper cysteine levels in the body can increase rate-limiting in glutathione synthesis, thereby increasing a favorable cellular environment, which in turn can increase performance. Glutathione is an important antioxidant responsible for reducing oxidative damage to cells. Two main pathways of oxidative damage are *proteolysis (the breakdown of protein into smaller peptides)* and *apoptosis (cell death).* By inhibiting these in skeletal muscles, performance can be increased.

L-Cysteine also plays a role in protein synthesis; however, not as in-depth as **L-Leucine**. Most of its benefits seen with L-Cysteine can be attributed to its increase in glutathione; however, as mentioned before, this effect is minimal when compared to NAC.

Benefits

- Aid drug induced cravings
- Antioxidant
- Cognitive enhancer
- Improve athleticism
- Reduce addiction symptoms

Possible Side Effects

- Nausea
- Constipation
- Vomiting
- Diarrhea
- Rashes
- Fever

Dosage

- For oral administration, the recommended dosage seems to be around 600 mg 2x a day for general antioxidant effects, although dosage may vary on symptoms/problems addressed.

WARNING

Please consult with a medical professional prior to administering any form of L-Cysteine. There is concern that L-Cysteine might cause bronchospasm in people with **asthma**. Those with **bleeding disorders**, L-Cysteine might increase the risk of bleeding as it has the ability to slow clotting. There is an additional concern that L-Cysteine can affect those with **liver problems** due to its interaction with enzymes in the liver. For those with Low Blood Pressure, L-Cysteine can lower blood pressure to unsafe levels.

L-DOPA *(L-3,4-dihydroxyphenylalanine)*

L-DOPA, which can also be known as *Levodopa*, is a precursor to the neurotransmitters *dopamine* and *epinephrine (adrenaline)*. It is also an amino acid that can be synthesized in the brain or kidneys with the amino acid L-Tyrosine. It can also stimulate the release of *Neurotrophic Factor (growth and survival of neurons)* in the nervous system. L-DOPA has been used for *Parkinson's Disease* and recently been sold as a testosterone booster.

The mechanism of L-DOPA seems to come from its ability to cross the blood-brain barrier; once in the brain, it can then be converted into dopamine by *aromatic amino acid decarboxylase*, increasing overall concentrations in the brain. However, this synthesized dopamine can be neurotoxic, creating *Reactive Oxygen Species* that can damage *mitochondria* and *DNA*, after its breakdown by the *monoamine oxidase* enzyme. **Antioxidants are not an** establis**hed way** to prevent treatment complications. Nonselective MAO inhibitors can cause dangerous interactions; specialist-prescribed selective MAO-B combinations require monitoring.

- This is extremely beneficial for those with *Parkinson's Disease* as *Parkinson's* is known to lower dopamine concentrations, which results in tremors, delayed movements, rigidity, and motor dysfunction. This happens since *dopamine* is a neurotransmitter that sends signals to other neurons and nerve cells in the brain and the body; this affects behavior and motor movement.
- Along with L-DOPA, administration of Vitamins B6, B12, folates, and antioxidants can be effective at improving symptoms as well as reducing L-DOPA's oxidative stress that arises after prolonged use.
- L-DOPA is not an established routine testosterone supplement; a proposed endocrine effect is that it accomplishes this through dopamine stimulation of *Gonadotropin-releasing Hormone (GnRH); GnRH* then stimulates the synthesis of *Luteinizing Hormone (LH)* at the *pituitary gland. LH* then stimulates *Leydig cells (located in the testes or ovaries)* to produce testosterone.

L-DOPA can affect those with *Polycystic Ovarian Syndrome* since it can stimulate male hormones which can potentially further reduce *Follicle Stimulating Hormone* secretion augmenting symptoms.

Proposed benefits (evidence varies)

- Aid w/Parkinson's symptoms
- Possible nootropic
- Increase dopamine
- Testosterone booster

Possible Side Effects

- Abnormal heartbeatt
- Behavior changes
- Nausea
- Vomiting
- Twitching

Dosages

- Parkinson's treatment uses individualized prescribed levodopa, often with carbidopa, and careful titration. *Mucuna powder and purifi*ed L-DOPA are not interchangeable self-treatment regimens.

WARNING

Seek medical advice before use. Long-term use of L-DOPA might cause toxicity and an increase in the inflammatory response. There is concern that L-DOPA may cause gastrointestinal bleeding at high doses in those with **intestinal disorders** or **ulcers**. If you are diagnosed with an over/underactive thyroid, L-DOPA may affect thyroid hormone secretions, which could make condition/symptoms worse.

L-Glutathione

Glutathione is an *endogenously (inside the body)* created antioxidant; it is used by every cell in the body. Oral bioavailability depends on digestion and formulation; some randomized human research has found increased body glutathione *stores.* Consuming **L-Cysteine** has been shown to increase glutathione concentrations in the body, but changes in stores do not establish all proposed clinical benefits of glutathione supplementation.

L-Glutathione can be absorbed in the intestines, with extracellular breakdown and resynthesis contributing to its handling; an absolute inability to influence intracellular stores is not supported. Once there, two enzymes *y-Glutamylcysteine Synthetase* and *glutathione synthetase* synthesize glutathione.

- **N-Acetylcysteine** is a well-studied precursor used to increase glutathione levels in the body, as it is able to successfully pass through the *First Pass Metabolism.*
- Although tissue transport and resynthesis vary, supplementation can increase glutathione levels in the blood, which seem to enhance *nitric oxide* boosters by slowing down *nitric oxide* breakdown.

When L-Glutathione is broken down, the molecules L-glutamic acid, L-cysteine, and glycine, may provide some benefit through individual interactions in the body. Regarding the antioxidant effect of glutathione, as mentioned above, some oral regimens have altered body stores, although clinical relevance remains uncertain. To increase glutathione levels in the body, reference **L-Cysteine**.

Glutathione produced in the body primes and prepares compounds for removal. It is non-enzymatic and regulates oxidative damage in the body. Its antioxidant mechanism comes from the enzyme *glutathione-S-transferases which prime* glutathione to donate an electron to oxidants that are able to receive electrons, stabilizing the free radical. *Glutathione peroxidase* is able to reduce *lipid peroxidation (where free radicals steal electrons from lipids in cells)* and turn hydrogen peroxide into water. These effects make up glutathione's antioxidant activity.

Proposed benefits (evidence varies)

- Oral trials can increase body stores; clinical benefits remain uncertain.

Possible Side Effects

Oral products can cause gastrointestinal effects; absence of reported harm does not establish long-term safety.

Dosages

Oral trials use varied doses; no standard effective regimen is established.

WARNING

Seek medical advice before use. Inhaled glutathione can provoke **asthma** symptoms. Oral, inhaled and injected preparations have different effects and risks; changes in blood stores do not establish treatment of serious disease. Injection or inhalation requires an appropriate medical indication and monitoring, not self-administration based on oral-supplement claims.

L-Leucine

L-Leucine is an essential *branched amino acid (BCAA)* highly involved in protein synthesis as well in the metabolism of *acetyl-CoA*. L-Leucine is a *ketogenic* amino acid that can be used in *ketogenic* diets. It provides the highest effect on protein synthesis out of all the *BCAAs* found in *BCAA* mixtures. L-Leucine can effectively activate *Mechanistic Target of Rapamycin Complex 1 (mTORc1)*, which is responsible for protein synthesis. The other two BCAAs, *L-Valine* and *Isoleucine*, can activate *mTOR* not as effectively though. Pure L-Leucine can be more effective than a mixture of BCAAs.

L-Leucine plays a role in cell signaling; one of its roles is being able to activate the amino acid-dependent *Rag GTPase heterodimers* in mTORc1; this is beneficial as it can more effectively activate protein synthesis than *insulin (insulin can promote protein synthesis via mRNA translation).*

- Simply put, no protein synthesis will happen unless amino acid levels are high enough. When these levels are met, the heterodimers mentioned above are switched to "on" or active, starting a chain reaction in *mTORc1* that activates protein synthesis.
    - The pathway of these mechanisms helps promote an *anabolic* environment that favors muscle tissue and it's enhanced when tied to exercise.

L-Leucine has been seen to increase *mitochondria* biogenesis by *SIRT1 (believed to be a key regulator in energy and metabolic regulation)* activation; through this, the *mitochondria* in muscle cells can increase resistance to exercise-induced fatigue. It can additionally increase overall energy expenditure since Leucine is a precursor to *acetyl-CoA*. *Acetyl-CoA* is used in the *Citric Acid Cycle* to release stored energy through the oxidation of *acetyl-CoA.*

- *CoA's thiol group forms thioesters such as acetyl-CoA;* thioesters have a high energy bond that happens to be *exergonic (dissipates to the surrounding area); the energy* released is then used for **ATP** and *GTP (GTP is used for RNA synthesis).*

This overall function of L-Leucine is stimulating and enhancing protein synthesis in the body, making it an *anabolic* supplement. Its function is rather complex and not fully understood.

Benefits

- Creates anabolic environment
- Promote protein synthesis
- Increases strength
- Increases lean muscle mass

Possible Side Effects

- Fatigue
- Loss of coordination

Dosages

- 2-5 gram a day with or without a meal

WARNING

Seek medical advice before use. L-Leucine may cause **lung failure** in people with **Amyotrophic Lateral Sclerosis/Lou Gehrig's disease. In people with Branched-chain Ketoaciduria**, L-Leucine may cause seizures and mental retardation at high doses. The use of L-Leucine in **chronic alcoholics** may lead to **liver disease,** which could also lead to **brain damage**. For diabetics, L-Leucine affects insulin release; medication may need to be modified. High doses of L-Leucine can increase **ammonia** in the body, which can become toxic and cause serious complications.

L-Lysine

L-Lysine is an essential amino acid since the body cannot create it. Like **L-Leucine,** it does play a role in the synthesis of protein; however, its effect is insignificant, and supplementation will not improve performance. It has been found to be involved in the development of **collagen** in connective tissue. It has also been marketed as a symptom reducer for the *herpes simplex virus.*

- L-Lysine has been implicated in the stabilization of collagen molecules as they can be found at the end of collagen chains.

Arginine is the main nutrient necessary for the *Herpes Simplex Virus,* as it promotes replication. Lysine is an analogue (similar in structure) to arginine. L-Lysine can then be taken up by the virus instead of Arginine by blocking off available receptors; this is beneficial as lysine is able to inhibit the virus from taking up arginine keeping it from replicating.

Arginine plays a role in maintaining normal immune system, protein synthesis, hormone production, and calcium regulation functions. An increase in dosage outside of normal dietary consumptions has not been seen to aid any particular function in the body. All research and data have not seen an increase in strength, muscle mass, or human growth hormone. Although L-Lysine is an important base foundation for all these functions, it seems that it plays more of a support role rather than a performance enhancer.

- There is talk about high doses of L-Lysine and various versions of Arginine being able to influence *growth hormone* levels. Even if that was the case, high doses of L-Lysine have been seen to cause gastrointestinal discomfort/problems. Combining Lysine and Arginine would be counter-productive as they are both *analogues* of each other; they have the same pathway and they both can cross the blood-brain barrier, so they would both compete for the same receptors.

One thing to note about its benefit: Combining L-Lysine with calcium supplements can increase calcium absorption in the intestines, which could aid those dealing with *osteoporosis.*

- Since Arginine can improve blood flow and vasodilation, this can mean an increase of nutrients to the bones, and since L-Lysine can increase the absorption of calcium in the intestines, if one is taken in the morning and the other at night, the time difference may allow for synergistic properties that can enhance mineral density in bones.

Benefits

- Reduce Herpes Simplex symptoms
- Osteoporosis aid
- Connective tissue support

Possible Side Effects

- Diarrhea
- Cramping

Dosages

- The recommended dosage is 30mg per 2.2 lbs/1 kg of body weight

WARNING

Please consult with a medical professional prior to administering any form of L-Lysine. There is concern that L-Lysine may cause diarrhea and cramps in children that have an **intolerance** to **lysinuric protein**. For those with Kidney disease, L-Lysine may worsen current conditions.

L-Methionine

L-Methionine is a sulfuric *essential amino acid* since the body cannot produce it. L-Methionine is important in many bodily functions. The body uses L-Methionine for tissue repair; it plays a role in skin pliability and growth of hair and nails. It can *chelate (bond to metal molecules for removal)* heavy metals which aids the removal process. Since it's a sulfuric compound, the Sulphur can protect cells from pollutants and enhance the absorption of **selenium** and **zinc**.

- Interestingly enough, it has also been used as an antidote for Acetaminophen and is effective at reducing oxidative damage due to Acetaminophen like **NAC**.

Methionine helps make up part of **S-adenosyl methionine (SAMe)** since it is made up of **methionine** attached to an **ATP** molecule. It can be found used all throughout the body. Its main function is to donate a methyl group to other molecules. This action provides support to all *metabolic* functions in the body. *Metabolic* functions can be thought of as food to fuel conversion for energy, elimination of wastes, repair and the rebuild of numerous tissues, etc.

- Methionine is also involved in **creatine** synthesis since the increased provisions of methionine are able to produce **SAMe** and **SAMe** is then able to donate the needed *methyl groups* for creatine synthesis in the body.

It is important to note that supplementation of L-Methionine is not recommended unless otherwise prescribed by a doctor. L-Methionine consumption is regularly high in the average person as it is found in high quantities in foods such as eggs, cheese, chicken, tuna, beef, bacon, nuts, and rice. There are quite a few other foods that contain L-Methionine. Although it is an essential amino acid, it is fairly easy to acquire your daily intake through basic food consumption. The danger lies in the extra supplementation of L-Methionine as it will not enhance any of these processes more than their current function. In fact, the exact opposite has been seen, like impaired growth while favoring the growth of cancerous tumors, as many cancerous tumors have a high need for methionine nutrition.

Benefits

- Regulates body functions

Possible Side Effects

These are related to self-administered supplementation not through food intake.

- Brain damage

Dosage

- 13-19 mg per 2.2 lbs/1 kg of body weight

WARNING

Please consult with a medical professional prior to administering any form of L-Methionine. There is concern that L-Methionine may promote the growth of some tumors. Those diagnosed with **acidosis** may be affected as L-Methionine can change the blood's acidity level. High doses of L-Methionine can increase levels of *homocysteine* which can cause hardening of the arteries increasing the risk of **heart disease** at high doses, as well as affect those with **methylenetetrahydrofolate reductase deficiency** since it might cause a build-up in *homocysteine* in the body.

L-Ornithine

L-Ornithine is an amino acid that does not have genetic coding; it is not involved in protein synthesis or assimilation, but it is a central part in the *Urea Cycle* since the ornithine concentrations are a rate limiter. It helps the body rid itself of *ammonia* which can affect endurance and performance, not to mention that *ammonia* is toxic to the body. It has additionally been implicated in indirectly promoting protein synthesis as well as increasing *Growth Hormone* release.

It is believed that Ornithine can indirectly promote protein synthesis in muscle tissue due to the fact that **L-Citrulline** can activate *mTORc1 (activates mRNA transcription of protein synthesis).* L-Citrulline is directly created from Ornithine and *carbamoyl phosphate.* L-Ornithine has been seen to promote *Growth Hormone* release by increasing *Ghrelin (a hormone released by the stomach) mRNA expression*. The *Ghrelin* receptor is also called the *growth-hormone secretagogue receptor (GHS-R), distinct from the growth-hormone receptor,* and plays a role in hunger and bodyweight outside of this scope; not much else is known in regard to performance-enhancing effects.

There are three amino acids involved in the *Urea Cycle,* **L-Arginine**, **L-Citrulline,** and L-Ornithine. L-Ornithine can also be produced in the body through the conversion of L-Arginine through the *arginase enzyme; this* conversion leaves *Urea as a byproduct* although no ammonia is used.

- In the *Urea Cycle,* L-Arginine is converted to L-Ornithine**,** which is converted to L-Citrulline; through these conversions the body is detoxified of ammonia.
    - *Carbamoyl phosphate is formed using ammonia and bicarbonate, then combines with ornithine* to form citrulline. Urea is released later when arginase converts arginine to ornithine. This cycle lowers toxic nitrogen by converting it into urea for excretion; urea is not the direct byproduct of ornithine becoming citrulline.
- After citrulline then converts to *argininosuccinate, argininosuccinate* can then be converted back to arginine, which would result in free arginine in the body, or re-enter the *Urea Cycle.*

Ammonia can increase during exercise through amino acid metabolism and breakdown of AMP in the purine nucleotide cycle. Excess ammonia may contribute to fatigue, but *nitric oxide* synthesis does not automatically produce ammonia because both molecules contain *nitrogen. Nitric oxide synthase converts arginine into nitric oxide and citrulline; it is a distinct reaction from urea formation and ammonia disposal.*

<u>Proposed benefits (evidence varies)</u>

- Increased endurance in prolonged events
- May promote muscle growth

<u>Possible Side Effects</u> (These were observed at high doses of 10 grams or more)

- Diarrhea
- Gastrointestinal discomfort

<u>Dosage</u>

- 2-6 grams a day, preferably the dose should be broken up to 1-2 grams before bed

WARNING

Seek medical advice before use. L-Ornithine may cause interaction with any **liver treatments** as the *Urea Cycle* takes place in the *mitochondria* of the liver's cells. For those with pituitary conditions, Ornithine can affect the current condition as it is believed that Ornithine can promote hormone production.

L-Taurine

L-Taurine is a conditionally essential amino acid, essential as there is a limit to the amount that can be synthesized in our cells, and uptake may come from blood cells. It can be found at high concentrations in meat. It is one of the few amino acids not involved in any function related to protein, yet it is found at high volumes in muscle tissue next to **glutamine**. It makes up over 50% of amino acids in cardiac tissue and has been seen as a neurotransmitter as it can be seen to excite neuron tissues. Taurine enters the cell through Taurine transporters that have similar properties to those of creatine and are found in the majority of mammalian tissue.

L-Taurine seems to act as a neurotransmitter by modulating the **GABA** and *glycine* signaling systems, both of which are neurotransmitters that seem to have performance-enhancing and antioxidant effects.

- It is believed to promote *anxiolytic (anti-anxiety)* effects through binding to *strychnine-sensitive glycine (implicated in anti-anxiety)* receptors, which can result in the lowering of stress/anxiety.
- It is also able to enhance memory and learning through enhancing energy changes in *GSK3Beta (involved in neuron cell development)* and *PI3K-AKT (involved in brain growth).*
- It can bind to iron and copper that could potentially become oxidative. This binding effect is known as a *chelating* effect allowing these heavy metals to become easily removed from the body.
- It may also be able to increase cell volume, by increasing the amount of water it can hold, a similar effect to that of creatine.
- It can also activate *angiogenesis (creation of new blood vessels)* via the *PI3K/AKT (involved in the cell cycle)* pathway thereby increasing cardiovascular health.
- It seems that a taurine deficiency leads to oxidative effects in the body, mainly from an increase in *Superoxide Production* due to dysfunctional *mitochondria*, as *mitochondria use* taurine to optimize oxidation of fatty acids. Taurine is also believed to increase *thioredoxin interacting protein (proteins involved in REDOX signaling) mRNA Expression*, which could be the reason as to the regulation of oxidation in the body.

REDOX stands for *Reduction* and *Oxidation*. This is crucial for antioxidant and metabolic effects in the body. *Oxidation* is where an atom loses an electron and oxidizes, while *Reduction* is where an atom gains an electron and becomes less oxidative. Generally forming a new compound, this is important in biological energy conversion of glucose, cell signaling to antioxidant activity in the body.

Benefits

- Performance enhancer
- Nootropic
- Antioxidant
- Anxiolytic

Possible Side Effects

Currently, there are no known common side effects due to a lack of data or anecdotal reports.

Dosages

- 1-3 grams a day has been deemed safe with no known side effects

WARNING

Please consult with a medical professional prior to administering any form of L-Taurine. There is concern that L-Taurine, at high doses, may worsen symptoms in those diagnosed with **bipolar disorder**. Children taking L-Taurine may also experience stunted growth.

<u>L-Theanine *(γ-glutamylethylamide)*</u>

L-Theanine is an amino acid that is seen as an *analogue (similar chemical structure)* of the amino acids **L-Glutamine** and **L-Glutamate**, and it can be found in green tea. This similarity allows it to be a competitive glutamate *antagonist (blocks or dampens the activation of the receptor),* which may be one of the reasons as to why it's so effective. Being that both glutamine and glutamate act as neurotransmitters, L-Theanine seems to have a positive effect on mental health and can be seen marketed as such.

L-Theanine has very similar chemical structures to that of **GABA**, **L-Glutamine**, and **L-Glutamate,** all of which are important neurotransmitters. Since L-Theanine is able to cross the blood barrier, this allows it to be able to promote the benefits of these neurotransmitters.

- L-Theanine is able to increase *dopamine, serotonin,* and GABA concentrations in the brain, which are known to promote relaxation type of effects.
    - L-Theanine has been seen to increase *mRNA expression* of GluN1 (glutamate receptor) *subunit (provides glycine binding sites; glycine binding has been seen as anxiolytic)* and *GluR1 glutamate receptor subunit (involved in synaptic plasticity).*
- Dopamine increase through L-Theanine consumption is minimal, lasting upwards of 40 minutes but can be negated through caffeine intake.
- L-Theanine also has affinities for *NMDA (involved in anxiety)* and *AMPA (involved in anxiety)* receptors; this may be due to its chemical structure being similar to glutamine thereby blocking glutamine's activation of these receptors.
- It has been seen to reduce levels of *corticosterone (stress hormone)* through positive modulation of the *hypothalamic-pituitary-adrenal axis (involved in the creation of corticosteroids such as corticosterone),* protecting the brain from stress-induced memory impairment.
- L-Theanine can increase alpha wave activity in the brain, which is seen in a relaxed but alert state of mind. This may be due to the modulation of the *hypothalamic-pituitary-adrenal axis.*

Overall, L-Theanine has various nootropic benefits and can be extremely beneficial as an additive for rigorous and stressful activities; in addition, its neuroprotective effects can be beneficial for "burn-outs."

<u>Benefits</u>

- Cognitive enhancer
- Alertness
- Non-sedative
- Promote relaxation
- Neuroprotector

<u>Possible Side Effects</u>

- Gastrointestinal discomfort
- Headaches
- Dizziness

<u>Dosages</u>

- For a relaxed but alert state of mind, 200-300 mg of L-Theanine (Synergistic with GABA)

<u>***WARNING***</u>

Please consult with a medical professional prior to administering any form of L-Theanine. Currently, L-Theanine has been seen as a safe supplement with very few side effects when taken at recommended dosages. Please be sure to consult with a medical professional prior to consuming L-Theanine.

L-Tyrosine *(4-hydroxyphenylalanine)*

L-Tyrosine is a nonessential amino acid that is one of the building blocks of protein; it is also a precursor to *catecholamines* like *dopamine* and *adrenaline*. It has also been implicated in the creation of some *thyroid hormones* which can promote *thyroxin* production. It has generally been used to offset fatigue and aid with stress-induced memory impairments and can be found marketed as such.

The mechanism and benefits of L-Tyrosine come from being the precursor to *catecholamines (Dopamine, Noradrenaline, and adrenaline).* It is important to note that when it comes to increasing concentrations of dopamine and other *catecholamines* within the body, L-Tyrosine cannot increase already normal levels; however, it has been found to be useful at sustaining normal levels in situations where concentrations may be altered such as stressful situations whether it may be mental or physical.

- *Phenylalanine* is converted into L-Tyrosine by the enzyme *phenylalanine hydroxylase.*
- L-Tyrosine can then become converted into **L-DOPA** through the enzyme *tyrosine hydroxylase.*
- L-DOPA is then converted into *dopamine* through *DOPA decarboxylase.*
- *Dopamine* then gets oxidized to *noradrenaline* through the enzyme *dopamine-beta-hydroxylase.*
- *Noradrenaline* converts to *adrenaline* by the enzyme *phenylethanolamine N-methyltransferase.*

Through this process, the body can continue, cycle, or stop at a certain conversion to utilize what's needed in the body. This is why supplementation of L-Tyrosine has been linked to the enhanced benefits. In stress-induced memory impairment, the concentration of *noradrenaline* has been observed to be negatively impacted; L-Tyrosine can effectively increase noradrenaline production to normal levels. The regulation of *dopamine* can also offset fatigue in the body thereby increasing performance, mainly as it can increase tolerance to heat stress, similar to **caffeine's** effect in pre-workouts as it is able to promote *Dopamine* concentrations.

- L-Tyrosine can also regulate the production of *thyroxin*, the hormone produced by the thyroid gland, which is responsible for *apoptosis (programmed cell death)* in those with low L-Tyrosine. This is believed to be of some aid for underactive thyroids; however, not enough is known to state this conclusively.

Benefits

- Stress reducer
- Cognitive enhancer
- Reduced fatigue

Possible Side Effects

- Nausea
- Dizziness
- Fatigue
- Heartburn
- Agitation
- Headache

Dosages

- 100-150mg per 2.2lb/1 kg of bodyweight

WARNING

Please consult with a medical professional prior to administering any form of L-Tyrosine. There is concern that L-Tyrosine may affect those diagnosed with **phenylketonuria** as it has been known to interact with phenylalanine levels. In those diagnosed with an overactive thyroid or Graves' Disease, Tyrosine can increase thyroid production of Thyroxine, which can affect thyroid function and/or treatment. L-tyrosine has been seen to affect **schizophrenia** as it can worsen symptoms due to its interaction with catecholamines.

Lecithin *(Phosphatidylcholine)*

Lecithin is a phospholipid and the most common phospholipid found in the body. A phospholipid is a major component in a cell membrane; phospholipids provide a protective sheath and support for the integrity of the cell. It is found at high concentrations in soybeans, eggs, and milk. There are various types of lecithin: phosphatidylserine, *phosphatidylcholine*, and *phosphatidylinositol.* Generally, benefits have been observed through *phosphatidylcholine* and it has been used as a supplement as it has a **choline** molecule.

Being that Lecithin is both *hydrophilic (attracted to water)* and *lipophilic (attracted to fat/oil),* it can allow both water and fats to pass through a cell membrane, which explains the abundance and importance of *phosphatidylcholine* in cell membranes. All cells found in the human body NEED to allow molecules to go in and out. *Phosphatidylcholine* affects energy production in the *mitochondria* as well. *Choline* is the precursor to *acetylcholine,* a neurotransmitter directly involved in memory and muscle control. *Acetylcholine* is an active compound that can enhance memory and can reduce stress while promoting feelings of relaxation.

Phosphatidylserine can be found in high concentrations in the brain and is needed for *myelin (a sheath that protects the nerve)* and nerve cell membranes. Once there, it is used in the cell structure as part of the cell's membrane. This structure layering can increase calcium and potassium ion enzyme sensitivity while reducing *acetylcholinesterase (increase the breakdown rate of the neurotransmitter acetylcholine)* receptors. This may promote higher *acetylcholine* concentrations. This effect may be more prominent in elderly individuals due to the decline in *acetylcholine* levels due to age. It also makes up part of the *myelin* structure as *myelin* is a lipid-rich sheath that provides protection and faster signaling processing.

Phosphatidylinositol has not been seen to promote any benefits, although it plays a role in extracellular signaling; there is a lack of information to conclusively state any benefit outside of **Inositol** benefits.

Lecithin has been used to break down fat at localized areas THROUGH INJECTION. Individuals should not proceed with this route; only a medically trained professional can perform a procedure like this with the least amount of adverse effects.

Benefits

- Possible nootropic
- Fat dispersion/breakdown

Possible Side Effects

- Diarrhea
- Nausea
- A feeling of fullness
- Abdominal pain

Dosages

- 500-2,000 mg a day

WARNING

Please consult with a medical professional prior to administering any form of Lecithin. There currently is a lack of data and research done with Lecithin; not much is known about its side effects or adverse effects.

Licorice Root *(Glycyrrhiza glabra)*

It can also be seen spelled as *liquorice,* which is the root of the *Glycyrrhiza glabra* plant. It has been used throughout the world as a sweetener. From China for traditional medicine and a flavor additive to even as far as an additive in US cigarettes, made to suppress cough and act as a bronchodilator *(decreases resistance in the lungs to take in more air).* One of the active compounds is glycyrrhizin; it is important to note that this molecule **CAN REDUCE TESTOSTERONE AND INCREASE CORTISOL**.

It's important to note that there are two versions of this supplement, one with *glycyrrhetinic acid (hydrolyzed from glycyrrhizin)* and one without, and can reduce testosterone as well.

- *Isoliquirtigenin,* an active compound, has been seen to prevent glutamate-induced death in neurons. It is also able to reduce *dopamine* release induced by drugs, which can possibly aid in reducing addiction.

Isoliquirtigenin has also been implicated in *endometrial* cancer toxicity with low toxicity to normal endometrial tissue. It accomplishes this through promoting expression of *caspase-7 (finalizes apoptosis/ cell death) & LC3B (involved in the removal of dysfunctional or unnecessary components)* protein.

- *Glycyrrhetinic acid* has been seen to provide neuroprotection via inhibition of *high-mobility group box 1 (regulates DNA organization and binding)* and *glutamate-*induced *NMDA (involved in memory and learning through synaptic plasticity) excitotoxicity cascade.*
- *Glabridin,* an active compound, is able to cross the blood-brain barrier; its *influx (inflow)* rate is higher than its *efflux (outflow)* rate. It is also able to normalize *acetylcholine* and *glutamate (neurotransmitters)* levels preventing over-excitation and reducing *excitotoxicity.* This can provide neuroprotection as it has been seen to reduce *malondialdehyde (a marker of oxidative stress).* This can provide *adrenal* support.

Glabridin has also been seen to be able to reduce neuroinflammation through inhibition of *NF-kB transcription (stimuli for inflammation signals).*

Proposed benefits (evidence varies)

- Antioxidant
- Anti-cancerous
- Adrenal support
- Cognitive support

Possible Side Effects

- High blood pressure
- Water retention
- Absent menstrual periods
- Weakness
- Tiredness
- Decreased sexual drive in men
- Paralysis
- Headache

Dosage

- 150-300 mg a day or 1800 mg a day without *glycyrrhizinic acid*

WARNING

Seek medical advice before use. Licorice Root may affect those with **heart disease** due to increased **blood pressure** and water retention. Those **with hormone sensitive conditions** such as, but not limited to, **endometriosis** and **ovarian and breast cancer**, symptoms or treatments may be affected due to its interaction with those cell lines and its phytoestrogen content. In those with low potassium levels, licorice root can lower potassium levels, further causing a safety concern as potassium is involved in muscle contractions. Glycyrrhizin can cause hypertension, low potassium, fluid retention and dangerous arrhythmias. Deglycyrrhizinated licorice differs; its benefits and risks are not interchangeable.

Longjack (*Eurycoma longifolia jack*)

Longjack, also known as Tongkat Ali and distinct from Panax or *Asian Ginseng,* is a bitter-tasting herb that is native to Malaysia, Indonesia and to a lesser extent, India, Thailand, and Vietnam. Well known in folk medicine, it was used to enhance fertility, sexuality as well as anti-aging through antioxidant and anti-inflammatory properties. Presently it has been sold as an *ergogenic aid (performance-enhancing)* in the fitness industry and a possible testosterone booster.

Longjack has been shown to activate *17-alpha-hydroxylase (CYP17) enzyme,* a key enzyme in the steroidogenic pathway that produces *androgens (steroid hormones)* in the body, mainly through the conversion of **pregnenolone** and *progesterone* to create more androstenedione and **DHEA**. It is believed through this activation it can increase free *testosterone* or at the very least *androgenic* hormones in the body since DHEA is the intermediate and basis for *androgenic* and *estrogenic* hormones. This pathway is not a reason to self-prescribe a*romatase inhibitors to block estrogen conversion; altering estrogen* can cause harm. IT IS IMPORTANT TO NOTE that there is very little research on this matter and standardization has been an issue, an increase in testosterone has been noted only on those with low testosterone.

- Through this effect, it has also been reported to aid *osteoporosis* caused by *androgen-deficiency,* as stated before, research is extremely limited on this matter as well.

Glycoproteins and *eurypeptide* are claimed to stimulate testosterone production and be pro-aphrodisiac, however, there have been no studies to support this. *9-hydroxycanthin-6-one, however,* seems to be the active compound that enhances sexuality in Longjack.

- *9-hydroxycanthin-6-one,* on the other hand, is able to improve erectile function by inhibiting internal Ca^{2+} *(Calcium Ions)* release and interfering with Ca^{2+} entry on the cell's surface. This would result in the anti-erectile effect as it forces penile contractions.

Proposed benefits (evidence varies)

- May increase fat-free mass
- May increase strength
- Aphrodisiac
- May increase testosterone
- Enhance erectile function
- Anxiolytic

Possible Side Effects

There currently are no known side effects for lack of research or anecdotal reports.

Dosage

- 150-400 mg daily for 5 weeks 100:1 extract
- 20-30 g of dry root

WARNING

Seek medical advice before use. Long-term safety and adverse effects remain uncertain. IT IS IMPORTANT TO NOTE many of the testosterone claims have been made by the patent under Longjack, meaning it could be company driven. Human evidence remains limited. Many products that claim to contain Longjack do not and have been found to have dangerous fillers. A patent or reputable label does not guarantee safety; do not replace prescribed care. EFSA did not establish safety of the standardized root extract it assessed in 2021 because of potential DNA damage; this assessment was extract-specific.

Lutein

Lutein is a *xanthophyll (yellow pigment)*, a subdivision of *carotenoids (organic pigments)*. It is found in high concentrations in the *macula (central part of the retina)* which is responsible for 3-color vision. It is studied in *Age-related Macular Degeneration (AMD)*, a cause of blindness in the elderly (65 and older). It seems to also block blue-light damage in the eye. Lutein is only synthesized in plants, as it plays a vital process in chlorophyll modulation, which in turn can be found at high concentrations in green leafy vegetables and carrots.

Reactive Oxygen Species (ROS) are natural byproducts from the result of oxygen consumption like that seen in photon light conversion to electrochemical signaling in the eye. Generally, *ROS* is used by the body to signal the immune system and other processes in the body. When *ROS* is produced to excess, you see free radical damage in the body.

- Lutein can serve as an effective free radical scavenger in the retina, thereby preventing any *oxidative stress* to the surrounding cells; Lutein can additionally assist in filtering *blue light* reducing *photochemical* damage to *photoreceptor (involved in converting light into signals in the visual system)* cells.
- Lutein absorbs blue light, contributing to retinal light filtering. Labora*tory* exposures can increase *Rea*ctive Oxygen Species in *retinal pigment epithelium, the cells involved in light absorption and the visual cycle,* and cause *apoptosis. Su*ch experiments do not establish injury from ordinary screen use or prevention by supplements; effects depend on intensity, wavelength and duration.

Photo-oxidative injury, pigment changes and scarring contribute to *AMD*. Lutein and Zeaxanthin can increase *macular pigment, prompting study* of antioxidant protection. AREDS2 supports their use within a specific formula for selected AMD patients; it is not a general cure or primary-prevention regimen. Cataract and AMD outcomes should not be treated as interchangeable.

Proposed benefits (evidence varies)

- May prevent Cataracts
- May prevent AMD
- Antioxidant

Possible Side Effects

- Yellowing of the skin

Dosage

- 6 mg to 10 mg a day, preferably with food.
- Food items such as spinach contain about 26 mg/cup

Lutein is absorbed in fat and transported through *lipoproteins (cholesterol), High-Density Lipoproteins (HDL)* having a 52% effectivity transport rate compared to *Low-Density Lipoproteins (LDL)* at 22%.

- Those with *Cystic Fibrosis* can benefit from this as *carotenoid* absorption is rather poor.

WARNING

Seek medical advice before use. It is important to note that many of these risk factors are affected by family genes. Lutein supplements have been seen to be safe to use daily in studies for up to 2 years.

Lycopene

Lycopene is a *carotene (red pigment),* one of the two classes in *carotenoids (organic pigment).* They are mainly found in red fruits and vegetables such as tomatoes, red carrots, and strawberries but can also be found in asparagus and parsley. Lycopene is seen as a strong antioxidant in the body as it is a highly effective free radical scavenger.

Increased Lycopene consumption has been seen to reduce *Prostate-Specific Antigen (PSA) levels; high PSA* levels can indicate prostate cancer or benign prostate growth. Additionally, increased Lycopene consumption has been seen to reduce *8-hydroxy-2'-deoxyguanosine (8-OHdG); this* is a *DNA Adduct or* in other words, a sign of oxidative stress that's associated with cancer.

- This may be due to its antioxidant effect, preventing *DNA* damage like double-strand breaks that can result in *TMPRSS2-ERG. TMPRSS2 (prostate specific androgen-response gene)* and *ERG (a gene involved in the cell cycle)* can accidentally be fused since they are both located on chromosome 21. This can result in *TMPRSS2-ERG* driven *ERG* overexpression that can stop cells from receiving *androgen* signaling causing the cells to become unregulated and unorganized like that seen in some prostate cancer growth.

This methodology seems to be applicable at reducing the risk of various cancers and is not only limited to prostate cancer. An increase of plasma Lycopene concentrations seems to be linked to a decrease of *8-OHdG* levels in breast cancer survivors.

Lycopene has also been seen to decrease *lipid peroxidation (where free radicals steal electrons from lipids, compromising the cell membrane)* and *Low-Density Lipoprotein (LDL) Oxidation.*

- *LDL* can attach to *macrophages (white blood cells),* causing them to become *foam cells. Foam cells accumulat*e in artery walls, which can lead to *atherosclerotic* build-up. (Or plaque build-up in the arteries for short.)

By decreasing oxidation, the risk of *foam cells* decreases as well, thereby decreasing the risk of cardiovascular disease.

Proposed benefits (evidence varies)

- Anti-cancerous
- Enhance cardiovascular health
- Reduce CDV risk

Possible Side Effects

- No current side effects have been seen

Dosages

- 20-30 mg a day

Like many *carotenoids,* it is absorbed in fat and transported through *lipoproteins (cholesterol), High-Density Lipoproteins (HDL)* having a 52% effectivity transport rate compared to *Low-Density Lipoproteins (LDL)* at 22%.

- Those with *Cystic Fibrosis* can benefit from this as *carotenoid* absorption is rather poor.

WARNING

Seek medical advice before use. Human trial evidence does not establish supplementation to prevent or treat prostate cancer; a general claim that it worsens or spreads prostate cancer is also unproven. Continue appropriate screening and prescribed treatment.

MCT (Medium-chain Triglyceride)

In nature, there are various types of *triglycerides (lipids or fats),* in either the division of saturated fats and unsaturated fats. Saturated fats are more likely to be solid at room temperature, while unsaturated fats tend to be liquid. It is important to know the distinction between healthy and non-healthy fats, while saturated fats seem to have less of a health effect than unsaturated fats; some do show promise like MCTs. With that being said, MCT is a saturated fat and can be primarily found in **coconut oil** and palm kernel oil, both can be solid/semi-solids at room temperature. They are also absorbed at a faster rate than *long-chain triglycerides* comparable to glucose uptake. Many people have used MCT Oil to lower cholesterol, burn fat and lower cardiovascular disease risk.

The main active components of MCT are *caproic acid, caprylic acid,* and *capric acid.*

- *Caproic acid* can be found in butter
- *Caprylic acid* can be found in coconut oil and palm kernel oil.
- *Capric acid* can be found in coconut oil, palm kernel oil, and butter.

Many marketed MCT Oil supplements are a mixture of *caprylic* and *capric acids* with a ratio anywhere from 4:1 to 1:1 *(Think of it this way, 4 grams of caprylic acid to 1 gram of capric acid, 1 serving ratio).*

Since MCTs are able to cross the *mitochondrial membrane,* they are able to more readily enter *beta-oxidation (the breakdown of fatty acids to make acetyl-CoA).* This results in excess of *acetyl-CoA,* which is used for energy production while another portion will go into *ketogenesis (ketone production).*

- *Acetyl-CoA* is also used in the *Citric/Krebs Cycle* to release stored energy; CoA's thiol group forms thioesters such as acetyl-CoA; thioesters have a high energy bond that happens to be *exergonic (dissipates to the surrounding area); the energy* released can be used for **ATP** and *GTP (GTP is used for RNA synthesis).* Rapid oxidation does not prevent energy storage or guarantee fat loss; MCT still contributes calories to tot*al* int*ake*.

MCT effects on blood lipids vary with the comparison oil and fatty-acid mix. A reliable LDL-lowering *or* cardiovascular-prevention benefit is not established; replacing unsaturated oils may be unfavorable. Weight-loss findings are modest and *depend on substitution within the diet.*

Benefits

- Burn fat
- Healthier fat

Possible Side Effects

- Stomach discomfort
- Intestinal gas
- Nausea
- Irritability
- Diarrhea

Dosages

- Unhealthy daily fats replaced by MCT Oil not to exceed 20-60 grams

WARNING

Seek medical advice before use. MCT Oil may worsen people with **Liver problems** since MCT Oil is metabolized primarily by the liver; this can cause complications for those who have liver problems like **cirrhosis**. In people with diabetes, MCT Oil can increase ketones in the body, which can affect insulin; this may affect **diabetics**.

MSM *(Methylsulfonylmethane)*

MSM is marketed as an anti-inflammatory, an antioxidant, and a joint health aid. It is an *organosulfur* compound, meaning that it contains sulfur. It can be found in many joint supplement products that contain **glucosamine** or **chondroitin** as they have synergistic properties.

MSM seems to be able to act like an *analgesic (pain reliever)* by blocking peripheral nerve conduction. It is additionally able to stop *NF-kB (protein involved in the cellular response to stress)* activation via inhibiting the inflammatory *cytokine (cell messengers/signaling) mRNA expression* of *Interleukin-6* and *Tumor Necrosis Factor Alpha*. The inhibition of *NF-kB* activation can reduce *Cyclooxygenase-2 (COX-2)* and *Inducible Nitric Oxide Synthase (iNOS)* expression as well. In short it has anti-inflammatory properties.

- *Cytokines* are messengers in the immune system responsible for inflammatory responses while *COX-2* and *iNOS* are involved in vasodilation and blood flow which play a role in inflammation; this mechanism has been seen as the cause of discomfort and pain in inflammatory conditions.

MSM can have joint protective effects due to its sulfur content since sulfur is a component that makes up cartilage, and sulfur is also able to trap water in cartilage. However, it is believed that these benefits come purely from a mechanism related to MSM rather than its sulfur component.

- It seems sulfur alone cannot replicate the same EFFECTIVITY as MSM, such as its anti-inflammatory or *analgesic* properties.

Since MSM is such a small molecule, it can move in and out of cells; this proves effective as it has been seen to have antioxidant effects on the mitochondria; this is beneficial since mitochondria are producers of Reactive *Oxygen Species (ROS)* thereby reducing oxidative damage and responses like *cytokines*. This can also result in a reduction of oxidative damage to cartilage which can further reduce inflammation.

- Generally, *ROS* is used by the body to signal the immune system and to keep a balance in the body. When *ROS* is produced to excess, you see free radical damage in the body.

Benefits

- Anti-inflammatory
- May aid osteoarthritis
- Antioxidant
- Pain reliever

Possible Side Effects

- Nausea
- Headache
- Diarrhea
- Affect allergy symptoms
- Insomnia
- Itching
- Fatigue

Dosages

- 500 mg to 3 grams 2x a day, 3 grams a day seems beneficial for osteoarthritis

WARNING

Please consult with a medical professional prior to administering any form of MSM. There is concern that MSM may affect those with circulatory problems such as varicose veins as it may increase swelling and pain when applied topically like a lotion. Currently, there are no known adverse side effects. Those allergic to sulfur, please be advised that MSM does contain sulfur and can possibly cause an allergic reaction.

Maca Root *(Lepidium meyenii)*

Maca is an edible root from *Brassicaceae* or mustard family, where vegetables such as broccoli, radishes, turnips, and kale can be found. The root itself is very similar to that of a radish or turnip with a starchy potato-like taste. Native to the Andes Mountains in Peru, it is found in altitudes around 13,500 feet above sea level and has a history of being used as an aphrodisiac. It can currently be seen marketed as such.

It has been used for both males and females and its proposed mechanisms include endocrine modulation, but consistent human hormonal effects are not established. It also does not seem to have any sort of *androgenic (male hormones)* effects on either men or women. It also does not seem to act through *phytoestrogen* and *phytosterol* as other plants; rather, effects on *pituitary* and *hypothalamus* function remain hypotheses.

- It has been proposed to influence hormone signaling through the *endocrine (hormones) system,* in women through promoting *Luteinizing Hormone* via the *negative feedback loop (the process of hormone balance)* which can increase overall *androgens (male hormones)* in the body. Although the depth of its mechanism is not fully known, there is data that contradicts *androgenic* effects.
- Maca seems to be more effective as it builds up in the persons' system over time.

Macaridine, however, may be the leading alkaloid in these effects. However, there currently is no agreed primary component as to which is the main active compound in Maca, *sterols, macamides, macaine,* and various others that have been isolated in recent years.

Maca has been seen to have antioxidant like effects throughout the body. In preclinical studies, Maca root has shown neuroprotective effects against oxidative damage; the mechanism as to how is not known. It has also been seen to promote *spermatogenesis (the reproduction of sperm cells)* independently of testosterone levels, mainly the initial stages where *mitosis (part of the cell cycle, division)* occurs.

It is important to note that Maca has very little research with inconclusive results. Although promise has been seen in regard to fertility and sex drive, not much is known in regard to its full mechanism and functions. More data and research are needed to fully understand its effect on the human body.

Proposed benefits (evidence varies)

- Increased sex drive
- Fertility
- Antioxidant
- May balance hormones

Possible Side Effects

- Due to lack of data not enough is known about common side effects

Dosages

- Human trials commonly use about 1.5–3.5 g/day of specific preparations; no standard effective dose is established.

WARNING

Seek medical advice before use. Maca may affect those individuals that are diagnosed with conditions including, but not limited to, **breast cancer, uterine cancer, ovarian cancer, endometriosis, uterine fibroids** as Maca Root may affect hormone levels. Goitrogen in Maca Root can affect thyroid function in those with **Impaired Thyroid Function**.

Magnesium

Magnesium is a chemical element found on the Periodic Table sitting at atomic number 12. It is in the Alkaline Earth Metal group where calcium and **Strontium** are found and is a shiny gray silver looking metal. Magnesium is an essential mineral that is extremely necessary to cells and to about 300 enzymes in the human body. One of the main functions being discussed here will be its function in energy metabolism, muscle function, and its importance in bones.

Magnesium is found in the body as magnesium ions; 25% is found in muscle tissue, and 60% is found in our bones. Magnesium plays an important role in **ATP** metabolism. *ATPase* are enzymes that are dependent on calcium and magnesium ions. *ATPase* breaks down **ATP** (Adenosine Triphosphate) into *ADP (Adenosine diphosphate).* This *dephosphorylation (removal of a phosphate group)* is used to create energy that all forms of life use.

- Magnesium also plays a role in muscle *contraction (this means any movement done by the muscle),* as it activates *ATPase* in the *sarcoplasmic reticulum (membrane in muscle cells)* to uptake *calcium ions* needed for movement.

This becomes extremely vital for any athletic event or sport; during strenuous activity, the body can break down and reform 529 oz of ATP, even though the body can only store 3 oz at one time.

- Increased magnesium intake, *if one is lacking*, can lead to increased metabolic efficiency, optimal heart rate, and oxygen consumption. The result is an increase in aerobic capabilities.

Bone surface magnesium levels increase with magnesium loading; the opposite happens in *hypomagnesemia (magnesium deficiency); magnesium* gets pulled from the surface of the bone in an attempt to balance magnesium levels.

- This can lead to matrix calcification retardation and *osteoblasts (bone growth cell)* inhibition, which is vital to bone development.

It's been noted that the average American only meets 60% of their daily magnesium requirement; even athletes are lacking, consuming around 70% of their daily amount. *Alcoholics, elderly* and *diabetics* have poor magnesium absorption, therefore, they are prone to magnesium deficiency.

Benefits

- Enhance aerobic function
- Improve energy
- May prevent osteoporosis

Possible Side Effects

- Gastrointestinal discomfort
- Nausea
- Vomiting
- Diarrhea

Dosages

- Adult RDA: men 400–420 mg/day, women 310–320 mg/day, by age. The 350 mg/day adult upper limit applies to supplemental/medication magnesium, not food.

WARNING

Seek medical advice before use. Magnesium may affect people with **bleeding disorders,** as magnesium has been seen to slow blood clotting. Magnesium intake may affect people with **kidney problems** as it can be difficult for the kidneys to filter the excess from the body.

Manganese

Manganese is a chemical element found on the Periodic Table sitting at atomic number 25. It is in the Transition Metal group where **Chromium** and **Iron** are found. It is a shiny metallic silver looking metal and has been used in stainless steel to prevent rusting and pitting. Manganese is important for normal body development and function. Manganese plays an important role as a *cofactor (helper molecules)* for enzymes that are used for neuronal health and function. As important a mineral as manganese is, overexposure to it can cause neurological disorders, especially in young developing children.

In general health, it is involved in the activation of the enzymes oxidoreductase, *transferase, hydrolase, lyase, isomerase,* and *ligase,* just to name a few.

- *Oxidoreductase*-involved in electron transport, used in mitochondria function or as an antioxidant.
- *Transferase*-involved in the transfer of groups like *acetyl-CoA (ref. Pantothenic acid).*
- *Hydrolase*-Breaks molecules apart into smaller molecules with water, *hydrolysis.*
- *Lyase*-Breaks molecules apart via other means than that of *hydrolysis* or *oxidation*
- *Isomerase*-Converts a molecule from one form to another, changing its properties.
- *Ligase*-Involved in creating a new molecule usually along with *hydrolase.*

Manganese is also a required *cofactor* to the antioxidant enzyme *Superoxide Dismutase (SOD).* Manganese in excess, however, can have a pro-oxidant effect. *SOD* is able to convert the *Superoxide Radical* (O_2^-) into oxygen and hydrogen peroxide. *Superoxide Radicals* are toxic and can cause oxidative damage in the body.

Mainly, as manganese is very dose-dependent, a deficiency may cause epilepsy, affect neural activity, and disrupt *homeostasis (stable equilibrium)* of manganese in the brain. On the other hand, overexposure can have extremely toxic and irreversible neurological damage and effects.

- There appears to be manganese in *glutamatergic neurons; these* are neurons that produce glutamate, which is a common neurotransmitter that can influence neurotransmission dependent on manganese levels.

Benefits

- Support normal body function

Possible Side Effects

- Loss of appetite
- Headaches
- Leg cramps
- Tremors
- Convulsions
- Hallucinations
- Irritability

Dosage

- Adult adequate intake: women 1.8 mg/day, men 2.3 mg/day. The adult upper limit is 11 mg/day total, not a target.

WARNING

Seek medical advice before use. Manganese may affect people with **liver disease** as it may be difficult to rid the body of manganese. If you have **iron deficient anemia**, you may absorb more manganese than normal. Those who work in industrial areas that contain industrial manganese are prone to severe side effects and possible manganese toxicity. Please track your general manganese consumption.

Mangosteen *(Garcinia mangostana)*

Mangosteen, also known as Purple Mangosteen, is part of the family *Clusiaceae.* It is a native plant of Southeast Asia but can also grow in tropical climates such as Florida or Puerto Rico; it has been part of Southeast Asia's traditional medicine for treating wounds, inflammation, and diarrhea. This particular plant has been advertised as having antioxidant, anti-inflammatory, anti-carcinogenic and anti-proliferative activities. These claims mainly come from its concentrations of *xanthones*, giving it the name "superfruit." The apoptosis and tumor-cell mechanisms below are mainly laboratory findings; mangosteen is not an established cancer treatment.

In Mangosteen there are various types of xanthones; alpha-mangostin has been seen to be able to induce *apoptosis (programmed cell death)*, be *anti-proliferative (reduce cell multiplication)* as well as shut off *mitochondrial* activity in tumor cells. This has led to inhibiting tumor expansion.

- One of the key factors that Mangosteen seems to possess is the ability to induce *apoptosis* through *cell cycle arrest*, in which cells that have damaged DNA undergo DNA repair or *apoptosis* in an effort to correct their DNA before *proliferation.*
- It is also able to disrupt mitochondrial function by removing *cytochrome c (crucial to the electron transport chain)* and placing it in the *cytoplasm (the material inside a living cell)* while additionally activating *caspase-3 & -9. Caspase-3* is involved in executing *apoptosis* while *caspase-9* initiates it.
- *Alpha-mangosteen* has been observed to decrease *BCL-2 (proteins that can cause or stop apoptosis)* while increasing *Bcl-2 associated X protein (BAX)* expression. *BAX* protein has been generally seen as *pro-apoptotic* and can activate *apoptosis* in a cell.

Alpha-mangostin has been seen to have anti-inflammatory effects through the reduction of signaling pathways related to inflammation which is beneficial to those with inflammatory conditions.

- *Alpha-mangostin* can decrease *Interleukin-1 (IL-1), Interleukin-8 (IL-8)* and *Tumor Necrosis Factor Alpha (TNF-a).* The primary function of these signalers is to respond to immune and inflammatory responses to any sort of insult to the body.

<u>Although *Alpha-mangostin* seems to be extremely toxic to cancer cells, which is a beneficial and positive effect, there still IS low toxicity to normal cells.</u>

<u>Proposed benefits (evidence varies)</u>

- Anti-inflammatory
- May be anti-cancerous

<u>Possible Side Effects</u>

For the lack of research and anecdotal data, there is not enough information to justify common side effects.

<u>Dosages</u>

For a lack of research and anecdotal data, there is not enough information to justify a set dosage.

- Adhere to product labels diligently.

WARNING

Seek medical advice before use. Mangosteen may affect those with bleeding disorders as it might affect clotting. There isn't sufficient data out there to state toxicity or long-term effects of using Mangosteen for long periods of time. What is known is that it can affect those undergoing **chemotherapy**; some claim long term use can lead to **lactic acidosis (lactic build up in the blood)**.

Melatonin

Melatonin is a derivative of the hormone serotonin; it is secreted in the *pineal gland* found in the brain, which primarily functions to regulate sleep-wake patterns. Many people have used melatonin to either regulate abnormal sleep patterns or as a sleep time aid. It has also been seen to have antioxidant effects, have relatively low short-term toxicity, but long-term safety is uncertain and supplement content can vary.

There are two melatonin receptors found in the *pineal gland, MT1* and *MT2.* Generally speaking, it is believed that there is another melatonin receptor *MT3 (may actually be just a protein)* which is believed only to be found in animals; however, it is believed that *MT3* is able to reduce eye pressure through its activation.

- This is beneficial when *Light* and *Dark* cycles are not in sync with our internal clock for sleep as it can reestablish balance in our *circadian rhythm* via *hypothalamic-pituitary-adrenal axis* modulation.
- *MT2* reduces *dopamine release; dopamine* plays a role in wakefulness by suppressing the release of melatonin. Melatonin is able to accomplish this by reducing *calcium influx; calcium ions* act as messengers for neurotransmitter release.
- *MT1* is able to inhibit *cortisol* secretion; high *cortisol* levels keep you awake throughout the day and should be low right before we sleep. When you wake up in the morning, *cortisol* is what gives you the discomfort that is associated with waking as it is meant to keep you awake.

Melatonin, in general, has been seen to lower *adrenaline* and *noradrenaline* secretion due to stress and lower circulating levels, mainly through the interaction of *noradrenaline* synthesis in the *Locus coeruleus* via the *MT1* receptor. *MT1 primarily* acts through cell suppression while MT2 acts through phase shifts in *circadian rhythms; this* may be through *reticular thalamus* via the *MT2* receptor.

Melatonin may offer antioxidant capabilities through neuroprotection effects. It is able to reduce harm from *beta-amyloid pigmentation,* which has been linked to *Age-related Macular Degeneration (AMD).* It can also offer protective effects in times of *reperfusion injury (blood return after a period of low oxygen)* through allowing an *Akt/mTOR pathway (responsible for various cell processes and the cell cycle).*

Proposed benefits (evidence varies)

- Sleep aid
- Antioxidant
- Neuroprotection

Possible Side Effects

- Headache
- Irritability
- Sleepiness
- Stomach cramps

Dosages

- Generally seen 3-5 mg 30 minutes prior to bed tend to facilitate and enhance the quality of sleep.

WARNING

Seek medical advice before use. Melatonin may affect those with **depression** as it may make symptoms worse. If you are prone to **seizures,** they can increase the likelihood of seizures. If you have a **bleeding disorder,** melatonin can make bleeding worse. Melatonin can increase blood sugar in **diabetics**, monitor sugar levels. Children need clinician-guided use; timing and dose depend on the sleep problem, and more is not necessarily better.

Methoxyisoflavone *(5-methyl-7-methoxyisoflavone)*

Methoxyisoflavone is a naturally occurring *phytoestrogen* that is found in soybeans that has a similar structure as **ipriflavone**. It has been marketed as a non-hormonal anabolic supplement. Its claims range from increases in muscle mass, strength, power as well as bone-building effects. Human research remains sparse. Its claimed benefits come from its ability to increase protein synthesis, lower cortisol levels which lead to quicker recovery time, while not affecting hormonal levels like *androgenic* substances like steroids. *A randomized training study did not improve body composition, strength or anabolic markers; proposed pathways do not establish bodybuilding benefit.*

- Through further research, you will find a lack of data out there on its mechanisms and its trials.

What you will find through hard in-depth search is as follows;

- It has *NF-kB* inhibition which means that it may have anti-inflammatory effects as well as anti-tumor effects.
  - The mechanism of these effects is not well known; the only thing that is currently known is that 2 parts of its atomic structure are needed to inhibit *TNF-Alpha (Tumor Necrosis Factor Alpha)* transcription activity of *NF-kB.*
- It also has the ability to affect *bone resorption,* which may be beneficial to those with *osteoporosis.*
  - It is able to accomplish this by inhibiting the *parathyroid hormone; this* hormone is responsible for putting more calcium back in the blood when levels are low; it accomplishes this by pulling calcium from bones.

Until further research is done, it still is a controversial supplement in the bodybuilding community. Some state it works, others claim it's a waste of money. The data out there is severely lacking and has not been replicated.

Proposed benefits (evidence varies)

- Anti-inflammatory
- May maintain bone structure

Possible Side Effects

For the lack of research and anecdotal data, there is not enough information to justify common side effects.

Dosages

- 200-400 mg 2x a day

IMPORTANT NOTE, supplementation can cause false positives for cannabinol use

WARNING

Seek medical advice before use. Methoxyisoflavone may affect those that will be going into surgery as it may affect **blood clotting** through its *NF-kB* interaction. It is also important to know that one of its metabolites can cause a false positive on a cannabinoid drug test. There is concern that methoxyisoflavone can affect those with **thyroid problems** due to its interaction with thyroid hormones.

Mexican Wild Yam *(Dioscorea villosa)*

Wild Yam, from the *Dioscoreaceae* family, is native to North America and is currently listed as an "at-risk" plant by the United States Plant Savers. It has been used in traditional Chinese and Ayurvedic medicine to treat female hormonal symptoms such as menopause, menstrual cramps, and some after-effects of childbirth. Its diosgenin can be used industrially to manufacture steroids, but the human body does not convert wild yam into **DHEA** or progesterone. Currently it has shown promise for inflammatory conditions such as *rheumatoid arthritis* and some hormonal conditions such as *endometriosis* and *adenomyosis*.

The most studied main bioactive compound in Wild Yam is *diosgenin*. In *rheumatoid arthritis, fibroblast-like synoviocytes* have lost their ability to hit their cell "checkpoint" if you will, to make sure they are functioning properly and only grow in adhesive surfaces. In preclinical models, *Diosgenin* has been seen to inhibit this growth and induce *apoptosis (programmed cell death)* in these cells through overexpression of *Prostaglandin E2 (involved in inflammation)* leading to *caspase-3 (executes apoptosis)* activation.

- *Fibroblast-like synoviocytes* are part of normal lining tissue in joints, as its main purpose is to sustain a stable environment while providing nutrients like **hyaluronic acid** for the *synovial fluid*, which is responsible for reducing friction in the joints.
- In *rheumatoid arthritis, fibroblast-like synoviocytes* increase rapidly creating an inflammatory environment, cascading down to further inflammatory responses leading to cartilage destruction from *cytokines (inflammatory messengers)* and *extracellular matrix* degrading enzymes.

In women, *matrix metalloproteinase 2 (MMP-2) & matrix metallopeptidase 9 (MMP-9) have been* implicated in *gynecological disorders* such as *adenomyosis, endometriosis,* and *endometrioid endometrial carcinoma.* Both *MMPs* are directly affected when *progesterone* and *estradiol* levels are affected, as in menopause or childbirth. All these disorders can be characterized as the abnormal growth and spread of cells that create lesions due to *MMPs'* abnormal function.

- *MMP-2 & MMP-9* are responsible for the breakdown of the *extracellular matrix,* molecules outside of the cell that provide support and structure to cells, in events like tissue remodeling.
- *Diosgenin* has been seen to inhibit various *MMP* enzymes to include *MMP-2* & *-9* via suppression of mRNA *expression* of *MMPs* and promoting *tissue inhibitor of metalloproteinase-2,* which can reduce the growth of *endometrial* lesions. This can aid *rheumatoid arthritis* as well.

Proposed benefits (evidence varies)

- May aid endometriosis
- Aid rheumatoid arthritis
- Anti-cancerous

Possible Side Effects

- Vomiting
- Nausea
- Diarrhea

Dosages

- 1-3 mg per 2.2 lbs/1 kg of bodyweight

WARNING

Seek medical advice before use. Mexican Wild Yam may affect those individuals that are diagnosed with conditions including, but not limited to, **breast cancer, uterine cancer, ovarian cancer, endometriosis, uterine fibroids** as Mexican wild Yam may affect hormone levels.

Milk Thistle *(Silybum marianum)*

Milk thistle, also referred to as wild artichoke and carduus marianus, comes from the *Asteraceae* family where daisies and sunflowers can be found. This flower has been traditionally used for liver disorders in Europe, the Middle East, China, and India. Extracts are generally derived from the seeds of milk thistle which contain the highest active compound amount of *silymarin.* It can be seen marketed as a detoxifying agent and for liver support. Trials in liver disease are inconsistent or limited; proposed antioxidant actions do not establish liver protection from supplements.

Silymarin is made up of seven *flavonolignans (silybin A & B, isosilybin A & B, silychristin A & B and silydianin). Taxifolin (a flavonoid)* can also be found but at near undetectable levels.

- The compound of *silymarin* has been seen to suppress *oxidative stress* damage seen in *ultraviolet radiation* while additionally inhibiting *xanthine oxidase,* an enzyme that generates *Reactive Oxygen Species (ROS).* This can be extremely beneficial to skin health via protecting cells.
    - *ROS* is used by the body to signal the immune system and maintain balance in the body. As *ROS* is produced to excess, that's when you see free radical damage in the body which can lead to *lipid oxidation. Lipid oxidation* is oxidative damage to cells, in which free radicals compromise the integrity of lipids in cell membranes.

This mechanism is not only seen on the skin but on other parts of the body, mainly the liver. It is able to reduce *ROS* created from the metabolism of toxic substances like alcohol and drugs.

- Additionally, *Silybin* has been seen to positively modulate the activation of *Activator protein 1 (involved in the cell cycle and apoptosis), STAT3 (involved in cell growth and apoptosis)* and *NF-kB (a protein involved in the cellular response to stress).* This is beneficial as an irregular function of these mechanisms has been implicated in cancers, chronic inflammatory diseases amongst others.
- *Silymarin* has been seen to promote and enhance DNA replication while also promoting cell protein synthesis, which can lead to cell repair after injury. Mainly through regulating *DNA topoisomerase 2-binding protein 1,* it is involved in DNA regulation and damage response.

Phytosomes (such as **lecithin***)* can increase the bioavailability of milk thistle into the body.

<u>Proposed benefits (evidence varies)</u>

- Liver protection
- Promote liver health
- Improve skin health
- Anti-inflammatory
- Antioxidant

<u>Possible side Effects</u>

- Nausea
- Diarrhea
- Intestinal Gas
- Bloating
- Loss of appetite
- Indigestion

<u>Dosages</u>

- Extract strength *and preparation vary. "µM" is a laboratory concentration, not an* extract-standardization percentage or usable oral dose; no standard liver-treatment regimen is established.

<u>***WARNING***</u>

Seek medical advice before use. Milk thistle may affect those allergic to plants in the Asteraceae family such as **ragweed**, **marigold**, etc. Milk thistle may lower blood sugar levels; diabetics should monitor their levels as medication may need to be adjusted. Exp*erimen*tal estrogenic activity warrants advice with hormone-sensitive conditions such as **endometriosis, uterine fibroids and breast, uterine or ovarian cancer; clinical relevance is uncertain.**

Momordica charantia **(Bitter Melon)**

The bitter melon is a vine-like plant similar to watermelons and squash and comes from the *Cucurbitaceae* family where the cucumber, pumpkins and zucchinis can be found. It is native to tropical and subtropical regions. It bears an edible bitter, green to yellowish fruit. The fruit itself has been consumed throughout Asia as food but has also been found to have been used in traditional medicine for diabetes and some stomach discomfort and can be seen marketed as such.

The active compounds in bitter melon are generally referred to as triterpenoids; four of the *triterpenoids* are able to stimulate *AMP-activated protein-kinase (can activate or deactivate the following enzymes); this* can result in lower blood sugar levels.

- *Charantin* contains *sitosterol* and *stigmasterol glucosides (glucose derivatives from plant glucose).* These molecules have been seen to enhance *glycolysis (turns glucose into energy molecules)* while providing *hypoglycemic (agents that lower blood glucose levels)* effects, mainly by increasing *glycogen synthase (converts glucose into glycogen)* enzyme levels while reducing *glycogen phosphorylase (rate-limiting in glycogenesis)* activity which could explain its antidiabetic effects.
    - Additionally, it seems that it may be able to lower orally consumed glucose levels without affecting insulin response. This is believed to be accomplished by enhancing glucose uptake, such as distribution to muscles, thereby enhancing glucose metabolism. It is also proposed that it prevents glucose absorption by inhibiting carbohydrate enzymes such as *alpha-glucosidase* and *alpha-amylase*.
- Bitter melon may be able to reduce fat through *cAMP (ATP derivative used in signaling)* activated *PKA (involved in lipid metabolism) apoptosis (programmed cell death)* of white fat cells which is the most prominent fat around your body, mainly due to its *linolenic acid* content.
- It is also able to reduce *PPAR Gamma 2; this* regulates fatty acid storage as well as *adipogenesis (fat cell creation).*

Changes in *HDL* cholesterol are not inherently a toxic effect, and experimental fat-cell apoptosis does not establish clinical harm or benefit. Human glucose-control evidence is limited; bitter melon must not replace diabetes treatment.

Proposed benefits (evidence varies)

- May lower body fat
- Anti-hyperglycemic

Possible Side Effects

- Stomach discomfort
- Hypoglycemia

Dosages

For a lack of data, no set dosage is known, adhere to label instructions.

WARNING

Seek medical advice before use. Bitter Melon may affect **diabetics** as it can lower blood sugar levels; medication may need to be modified. If you have **glucose-6-phosphate dehydrogenase G6PD**, bitter melon can cause anemia. Toxicity mentioned above can affect **fertility** as well as **lymphocytes** which is of concern for those with immunodeficient conditions. Potential glucose-lowering and medicine interactions need individual review; HDL increases alone are not evidence of **cardiovascular** toxicity.

Mucuna pruriens

Mucuna Pruriens is a legume of the *Fabaceae* family where beans and the soybean can be found, and can also be called velvet bean or cowitch. It is native to Africa as well as some tropical areas in Asia. The beans are protected by a lining of hairs that contain *serotonin* and *mucanin* which cause uncontrollable itching. The main active component is **L-DOPA**, a precursor to the neurotransmitters *dopamine* and *epinephrine (adrenaline).* Additionally, it is able to stimulate a *Neurotrophic Factor (growth and survival of neurons)* release in the nervous system. L-DOPA has been used for *Parkinson's Disease* and recently been sold as a testosterone booster.

The mechanism of L-DOPA seems to come from its ability to cross the blood-brain barrier, as dopamine is not able to. Once in the brain, it is able to be converted into dopamine, thus, increasing dopamine concentrations in the brain.

- This is extremely beneficial for those with *Parkinson's Disease* as *Parkinson's* is known to lower dopamine concentrations, which results in tremors, delayed movements, rigidity, and motor dysfunction.
    - Dopamine is a neurotransmitter that sends signals to other neurons and nerve cells in the brain and the body; this affects behavior and motor movement.

L-DOPA may increase testosterone in the body. It accomplishes this through dopamine stimulation of *Gonadotropin-releasing Hormone (GnRH); GnRH* then stimulates the synthesis of *Luteinizing Hormone (LH)* at the *pituitary gland. LH* then stimulates *Leydig cells (located in the testes or ovaries)* to produce testosterone.

- Studies are extremely limited to conclusively state its effect on boosting testosterone.

Administration of Vitamins B6, B12, *folates,* and antioxidants can be effective at improving symptoms as well as reducing L-DOPA's oxidative stress that arises after prolonged use; this is not established by choosing Mucuna instead of prescribed levodopa, and co-supplements require medication review.

Proposed benefits (evidence varies)

- Aid Parkinson's symptoms
- May boost testosterone
- Possible nootropic
- Aid with low dopamine levels

Possible Side Effects

- Abnormal heart beat
- Behavior changes
- Nausea
- Vomiting
- Low blood pressure
- Twitching

Dosages

- L-DOPA content varies widely across Mucuna products; a fixed powder weight does not give a predictable drug dose.
- Parkinson's treatment requires individualized prescribing and monitoring, often levodopa with carbidopa. Do not substitute a 500–1,000 mg supplement regimen for prescribed care.

WARNING

Seek medical advice before use. Long-term use of L-DOPA may cause toxicity and increased inflammatory response. There is concern that L-DOPA may cause gastrointestinal bleeding at high doses if you have **intestinal disorders**. If you are diagnosed with an overactive or underactive thyroid, L-DOPA may affect thyroid hormone secretions, which could make the condition/symptoms worse.

N-acetyl cysteine/N-acetyl-L-cysteine (NAC)

L-Cysteine and N-acetyl cysteine are generally grouped together as the same. These two supplements, although similar, are **NOT** interchangeable. Granted, the benefits are similar; however, NAC enhances and gives more notable benefits. The reason for this is that the *First Pass Metabolism (drug breakdown by the body)* of L-Cysteine is broken down and distribution is minimized. The *acetylated* cysteine diminishes this breakdown effect, increasing absorption in the intestines and allowing it to cross the blood-brain barrier.

NAC has been seen as an antioxidant in the sense that it supports **glutathione** synthesis. Glutathione is an important antioxidant responsible for reducing oxidative damage to cells. Increased NAC levels can result in an increase of glutathione levels, which can lead to lean muscle mass since it can protect cells from oxidant damage. Maintaining proper cysteine levels in the body can increase the *rate-limiting (the amount that can be made)* in glutathione synthesis, thereby increasing a favorable cellular environment, which in turn can improve general health.

- Two main pathways of oxidative damage are *proteolysis (the breakdown of protein into smaller peptides)* and *apoptosis (cell death)*. By inhibiting these two in skeletal muscles, performance can be increased.
- By inhibiting *proteolysis* and *apoptosis,* it can also benefit various other areas in the body by lending the same protective effects, such as the liver. NAC is an established medical treatment for acetaminophen poisoning; routine protection from alcohol or improved muscle gain is not established.
- It can additionally provide anti-inflammatory effects by inhibiting the separation of *IkB* from *NF-kB (where inflammation begins)* reducing the inflammatory signaling cascade.

NAC is able to cross the blood-brain barrier, thereby increasing neural glutathione concentrations, providing cerebral antioxidant effects. NAC reduces glutamate levels/stimulation in the brain. High glutamate levels can be seen in addictions such as drug use or obsessive-compulsive disorder. Oxidative damage appears to increase addiction severity; glutathione can provide benefits to this.

- NAC can additionally protect neurons from *dopamine*-induced *apoptosis* while also enhancing *cytochrome c (crucial to the electron transport chain)* in mitochondria reducing oxidative damage.

Proposed benefits (evidence varies)

- Reduce drug-induced cravings
- Antioxidant
- Hepatic protectant
- Aid athletic performance
- Aid neurologic function
- Anti-inflammatory

Possible Side Effects

- Nausea
- Constipation
- Diarrhea
- Vomiting

Dosage

- 600 mg 2x a day for general antioxidant effects, dosage may vary on problem/illness.

WARNING

Seek medical advice before use. Inhaled NAC can cause bronchospasm, particularly with **asthma; oral and intravenous risks differ**. Review bleeding risks and medicines, inc**luding nitroglycerin**, which can increase headache and low blood pressure. Acetaminophen poisoning needs urgent medical treatment, not a routine supplement dose.

Niacin

Niacin, also known as *nicotinic acid,* is part of the Vitamin B_3 complex that includes *nicotinamide, nicotinamide riboside,* and *nicotinic acid.* It is a water-soluble vitamin that has been used for improving vascularity and normalizing and correcting cholesterol/lipid levels. It plays a key role in *REDOX* in cells.

- *REDOX* stands for *Reduction* and *Oxidation.* This is crucial for antioxidant and metabolic effects in the body. *Oxidation* is where an atom loses an electron and oxidizes, while *Reduction* is where an atom gains an electron and becomes less oxidative. Generally forming a new compound, this is important in the biological energy conversion of glucose to antioxidant activity in the body.

Niacin, in its normal state, has been used by athletes for its flushing and vasodilation effect. The increased blood flow seems to come from an increase of *bilirubin (blood antioxidant)* indirectly due to the niacin supplementation. It is believed that *bilirubin* can enhance *nitric oxide effect on* the body for longer periods of time. While the flush seems to come from niacin's effect to widen the vessels in the skin.

It is able to improve lipid profile and normalization via inhibiting the uptake of *High-density Lipoprotein.* Mainly through preventing degradation of *apoprotein A*, this protein plays a role in removing fats from white blood cells and artery walls and returning it back to the liver. Niacin can indirectly inhibit the synthesis of fatty acids through decreasing *Protein Kinase A (involved in converting triglycerides into fatty acids)* activity. This can reduce *Very Low-density Lipoprotein (vLDL)* as fatty acids are needed to create *vLDL.* It additionally directly inhibits *Diacylglycerol acyltransferase 2,* an enzyme needed for the creation of *triglycerides* in the liver.

- This can reduce the risk of *foam cells; foam cells* are caused by *oxidized LDL cholesterol* attaching to white blood cells. *Foam cells accumulat*e in artery walls, which can lead to *atherosclerotic* build-up.

Proposed benefits (evidence varies)

- May promote vascularity
- Lower cholesterol
- Antioxidant

Possible Side Effects

- Flushing
- Stomach discomfort
- Increased blood pressure
- Diarrhea
- Redness of skin
- Paresthesia
- Itching
- Nausea

Dosages

- Adult RDA: women 14 mg NE/day, men 16 mg NE/day. The adult supplemental upper limit is 35 mg/day. Amounts such as 200–500 mg are pharmacological, not nutritional doses. *Immediate-, sustained-* and *extended-release* products differ; reduced flushing does not eliminate liver, glucose or other risks. Use high-dose niacin only with medical monitoring.

WARNING

Seek medical advice before use. Large doses of Niacin can cause irregular heartbeat as well as **lower blood pressure**. Niacin can possibly increase sugar levels in **diabetics**. Niacin may complicate **gallbladder disease**. If you have liver disease, niacin may increase liver damage. In those with intestinal/stomach ulcers, niacin may worsen symptoms.

<u>Noni *(Morinda citrifolia)*</u>

Noni comes from the *Rubiaceae* family where the coffee tree can be found. It is native to Asia and in tropical regions as well. It has been used in traditional Polynesian medicine. This fruit can be found advertised as a miracle product throughout the health and fitness industry. The marketed health benefits have been claimed to be anti -bacterial, -fungal, -viral, -tumor, -inflammatory, -parasitic, and an immune booster. Human clinical benefit for these advertised uses is unestablished; isolated-compound mechanisms do not show that noni juice treats disease.

Antibacterial activities seem to be derived from *phenolic compounds: ferulic acid, hydroxycinnamic acid, methylparaben, methyl ferulate, methyl 4-hydroxycinnamate,* and *15-dimethylmorindol*.

- *Ferulic Acid* has been seen as an antioxidant as it is reactive with free radicals responsible for cellular damage like *Reactive Oxygen Species (ROS).* It has also been seen to reduce *thromboxane* B_2 production. *Thromboxane 2* can facilitate platelet aggregation as well as constrict blood vessels. *Ferulic Acid* can also reduce oxidative stress on the skin when applied topically, thereby reducing *thymine dimers (these can cause cancer) on the skin.*
- *Methyl Ferulate* is looked at as a possible anti-inflammatory as it inhibits the *MAPK (Mitogen-activated Protein Kinase) Pathway*, the cell signaling from the surface of the cell to the DNA inside the nucleus. *MAPK* is directly responsible for pro-inflammatory *cytokines (cell signaling/messengers).*
    - *MAPK* mutations are a factor in the growth of tumors; methyl *ferulate* may reduce the growth of tumors by inhibiting the function of *MAPK pathways,* including mutated ones.
- *Methylparaben* can be seen to stimulate *cell-mediated immunity (immune response without antibodies)* as well as stimulate a histamine response. It can inhibit the growth of molds and various fungal spores, and for that reason, it can be found in food packaging.
- *15-Dimethylmorindol* has been seen to suppress the malaria parasite, although its interactions aren't fully known; it was able to reduce parasites' ability to survive.
- *4-hydroxycinnamate* has been seen to promote *caspase-3 (involved in executing apoptosis/cell death)* in cancer cells.

<u>Proposed benefits (evidence varies)</u>

- Antioxidant
- May be anti-carcinogenic
- Anti-parasitic
- Anti-fungal
- Anti-viral
- Anti-inflammatory
- Immune booster
- Antibacterial

<u>Possible Side Effects</u>

- Gastrointestinal discomfort
- Diarrhea
- Heavy menstruation
- Acne
- Bad breath
- Headaches

<u>Dosages</u>

- Dosage may be used as follows: 50-750 mL of noni juice a day or 500 mg of extract a day.

<u>***WARNING***</u>

Seek medical advice before use. High potassium can be dangerous with **kidney disease** or potassium-raising m**edicines**. Liver injury has been reported. Review clotting-related medicines; with diabetes, check whether juice contains added sweetener. Neither a natural source nor an antioxidant mechanism establishes safety.

Ornithine Alpha-Ketoglutarate (OKG)

OKG is a salt; it is created by combining two Ornithine molecules with one Alpha-Ketoglutarate molecule reaching a 1:2 *stoichiometric (The basic measure of the amount needed to get a given product amount or reaction)* ratio. OKG has been seen to preserve muscle mass by decreasing muscle catabolism while increasing protein synthesis.

In the *Urea Cycle, arginase converts arginine* to **ornithine** and releases urea. Ornithine and carbamoyl phosphate form citrulline; **citrulline** returns to arginine through argininosuccinate. Ammonia supplies nitrogen through carbamoyl phosphate. The cycle disposes of toxic nitrogen as less-toxic urea.

- When Alpha-Ketoglutarate is added, it can further buffer the body of ammonia; alpha-ketoglutarate and ammonia can be used as intermediates of **glutamine** metabolism, thereby combining to form glutamine.
- It is also believed that Ornithine can indirectly promote protein synthesis in muscle tissue since **L-Citrulline** can activate *mTORc1 (can activate mRNA transcription of protein synthesis).*

Exercise can increase ammonia through amino acid metabolism and AMP breakdown, rather than as a direct byproduct of *nitric oxide* synthesis. Excess ammonia may contribute to fatigue. Alpha-ketoglutarate contributes carbon skeletons to glutamate and glutamine synthesis, which can incorporate nitrogen; this mechanism does not establish an exercise benefit from OKG in healthy athletes.

OKG as a whole provides benefits that the two separate compounds cannot; it is able to preserve muscle mass by maintaining *ribosome (they are extremely biased against amino acids)* concentrations. *Ribosomes* are extremely effective at incorporating amino acids into protein; without ribosomes, there would be no protein synthesis, meaning no muscle repair/maintenance. A high concentration of *ribosomes* is a good measure of protein synthesis capabilities.

- OKG can provide enough carbon skeletons to compensate for lack of glutamine and other amino acids that are typically needed to provide repair and maintenance of muscle tissues, in cases like trauma, injury or day to day bodily process. These carbon skeletons can allow for the creation of these amino acids like glutamine, allowing OKG to sustain muscle tissue.

Benefits

- Preserve muscle mass
- Improve performance
- Reduce Fatigue
- Balance Nitrogen environment

Possible Side Effects

There have been no known reported side effects

Dosages

- 10-20 grams a day preferably with carbohydrates

WARNING

Seek medical advice before use. OKG may affect people with **Ornithine-delta-aminotransferase**, a rare genetic but serious condition which can result in progressive blinding. Currently, there have been no documented adverse reactions or interactions.

Olive Leaf

Olive leaf is exactly as the name states, the leaf from the Olive tree. Its extracts are said to have higher concentrations of *phenolic* compounds, although there is insufficient scientific evidence to conclusively state its health benefits; it has been seen to improve cardiovascular health by inducing anti-inflammatory and anti-atherosclerotic effects while also lowering blood pressure and cholesterol.

Some of the main active compounds found in the olive leaf are: *secoiridoid oleuropein, hydroxytyrosol, tyrosol, caffeic acid,* and *luteolin-7-glucoside.*

- *Secoiridoid oleuropein* may be responsible for Olive Leaf's hypotensive effect as it is able to act as an antagonist/inhibitor of the calcium ions channel. Calcium plays a part in blood pressure regulation and muscle contraction. Another way it may lower blood pressure is through *Angiotensin-converting-enzyme (controls blood pressure)* inhibition, which ultimately lowers arteriolar resistance, decreasing cardiac output while increasing venous capacity.
- *Hydroxytyrosol* and *tyrosol* may lower *LDL (Low-density Lipoprotein)* oxidation indirectly by interacting/suppressing *endothelium (the interior surface layer of vessels, i.e., blood/vascular cells)* inflammatory response. This can reduce *foam cells; foam cells* are caused by *oxidized LDL cholesterol* attaching to white blood cells; this can reduce cell adhesion and plaque build-up.
- *Luteolin-7-glucoside* has been seen to either reduce the absorption or the rate of absorption of carbohydrates while also having a mild inhibiting effect on carbohydrate enzymes. This may help reduce sugar levels in the body.
- *Caffeic Acid* alongside *oleuropein* has been seen to have anti -bacterial/-viral effects, although the general compounds of Olive Leaf tend to be more efficient at inhibiting bacterial and viral growth than these two components alone.

**There's a claimed "detoxification reaction" reported by many users, in which your kidneys, liver, etc. become overwhelmed with dead pathogens as a result of the Olive Leaf. For the "Detoxification Reaction," side effects reference Possible Side Effect below.*

Benefits

- Lower blood pressure
- Cardiovascular health aid
- Lower LDL Cholesterol
- Antioxidant
- Lower blood sugar
- Antibacterial
- Antiviral

Possible Side Effects (Reference note above)

- Headaches
- Fatigue
- Body aches
- Skin rash
- Diarrhea

Dosage

- For general aid with cholesterol, the consumption of olive oil products may provide benefits.
- 500-1000 mg a day of an extract

WARNING

Please consult with a medical professional prior to administering any form of Olive Leaf Extract. There is concern that Olive Leaf Extract may affect those diagnosed with **diabetes** as it may affect blood sugar levels; please monitor levels accordingly. It is possible to be allergic to Olive Leaf if one is allergic to Olive pollen; please talk to a medical professional if you have allergies towards Olive pollen.

Omega-3

It is important to note that there are 3 types of omega-3 fatty acids: docosahexaenoic *(DHA)* and *eicosapentaenoic (EPA) acids, and Alpha-Linolenic Acid (ALA), ALA is essenti*al, while limited conversion of ALA can produce EPA and DHA. Since *ALA* has a poor bioconversion towards *EPA* and *DHA,* all focus and studies were aimed towards *EPA* and *DHA* supplementation, since it was found that they had protective effects against *cardiovascular disease.* Ordinary supplements do not reliably prevent *cardiovascular* events; prescription icosapent ethyl has benefit in selected high-risk patients, and high doses can increase atrial-fibrillation risk. ALA, EPA and DHA are not clinically interchangeable; more ALA does not guarantee equivalent effects.

Docosahexaenoic (DHA) and *Eicosapentaenoic (EPA) Acids are* generally derived from **fish oil** or **krill oil**, *with no univers*al clinical superiority established for **krill oil**. *DHA* seems to have nootropic effect that is further enhanced by EPA's ability to increase blood flow to the brain; this may aid those diagnosed with cognitive problems. It is important to note, those who regularly DO NOT eat fish as part of their diet, *DHA* may increase cognitive functions; for those that DO, benefits *may not be seen from this supplement*.

It's also been seen to reduce inflammation through the suppression of *phosphoinositide 3-kinase,* a signal transducer involved in inflammation. It can also reduce inflammation through modifying the production of eicosanoids; eicosanoids are messengers that are released after **arachidonic acids** are broken down; these messengers can either inhibit or cause inflammation.

- This may reduce plaque build-up in the arteries as it prevents *endothelium (the interior surface layer of vessels, i.e., blood/vascular cells)* inflammation, increasing cardiovascular health. This may be beneficial at reducing joint inflammation especially those with osteoarthritis.

Omega-3 has been seen to have the ability to enhance protein synthesis as well as affect *myostatin* (*GDF-8*) abilities. *Myostatin*, in short, is a growth factor; simply put, myokine is a protein that's released by *myocytes* that directly affects the muscle cells' ability to signal the inhibition of *myogenesis* (creation of muscle tissue).

Proposed benefits (evidence varies)

- Aid with arthritis
- Cardiovascular health aid
- May promote muscle growth
- Nootropic
- Promote joint health
- Anti-inflammatory
- Neuro-protectant

Possible Side Effects

- Loose stools
- Nausea
- Fishy breath (if fish oils are the supplement)
- Upset stomach

Dosage

- 1 gram a day to include EPA & DHA generally 735 mg for EPA & 265 mg for DHA
- 1-3 grams a day for ALA from **flax** or **chia seeds** for a vegan alternative

WARNING

Seek medical advice before use. This brand of Omega-3 may affect those with **Seafood/Shellfish allergies** if it's derived from fish products. In people with bleeding problems, omega-3 can increase the risk of bleeding. Omega-3 may make blood pressure too low for people with **low blood pressure.**

Omega-6

Omega-6 like **omega-3** is a *polyunsaturated fatty acid,* and essential linoleic acid cannot be synthesized, though other omega-6 fatty acids can be made from it. One of the main purposes of omega-6 fatty acids is to modulate proper inflammatory responses in the body, especially after physical activity. This can attenuate cell damage while promoting cell repair; this is especially beneficial to athletes as it can promote muscle growth and repair.

Arachidonic Acid (AA) is a commonly known omega-6 fatty acid responsible for this action. Some AA-derived mediators p*romote clotting or inflammation; this does not mean ordinary dietary linoleic acid increases these disease risks.* (Reference Arachidonic Acid for more information)

- AA accomplishes this through its breakdown; what's leftover are its metabolites, *leukotriene 4* and *thromboxane 2.*
    - Leukotrienes are messengers responsible for inflammatory responses. Leukocytes (White-blood cells) create leukotrienes from the oxidation of AA.
- *Thromboxane 2* can facilitate platelet aggregation as well as be a vasoconstrictor *(constriction of blood vessels and vascular muscle cells)*

Similar to the same situation as omega-9 fatty acids, the American and European Diet already includes more than enough omega-6 fatty acid intake that supplementation may harm one's health.

- No universal optimal omega-6:omega-3 ratio is established. Dietary patterns can have widely different rati**os, which alone do not show whether either fatty acid intake is adequate or** harmful.
    - The ratio compares amounts of omega-6 and omega-3 consumed; for example, 4:1 means four parts omega-6 per one part omega-3. It is a description of intake, not a clinical prescription.
- A ratio above 4:1 does not itself establish disease causation. Evidence supports replacing saturated fats with polyunsaturated fats, including linoleic acid, within a balanced diet.

Proposed benefits (evidence varies)

- Modulate inflammation
- Reduce cell damage
- May promote muscle growth

Possible Side Effects

Normal dietary intake is generally tolerated. Extra supplementation is usually unnecessary; theoretical inflammatory pathways do not establish harm from ordinary linoleic-acid-rich foods.

Dosage

- Meet essential fatty-acid needs through an appropriate diet. There is no established requirement to limit omega-6 to four times omega-3 or to pursue a 1:1 ratio. Food sources and the nutrients being replaced matter.

WARNING

Seek medical advice before use. Most diets already supply adequate linoleic acid. Dietary omega-6 is not inherently harmful; consider overall diet, energy balance and appropriate omega-3 intake instead of diagnosing risk from the ratio alone.

Omega-9

An omega-9 fatty acid is an *unsaturated fatty acid,* more specifically a *monounsaturated fat. Saturated and unsaturated dietary fats* both provide about 9 kcal per *gram.* The main difference between the two: one is saturated with a bunch of hydrogen atoms, while the other is not, hence the term saturated or unsaturated. Omega-9 fatty acids are considered non-essential fatty acids as they can be created in our body so as long as your diet consists of enough *unsaturated fats.* For the most part, supplementation is not necessary; the majority of people already consume enough omega-6 and omega-9 fatty acids in their diet. Omega-9 fatty acids are found in vegetable oils like olive oil. Omega-6 fatty acids can also be found in vegetable oils and in most meats.

Omega-9 fatty acids can regulate inflammation and lower cholesterol levels as well as aid in increasing energy levels.

- Intake of omega-9 fatty acids has been shown to reduce Total Cholesterol levels; this includes *Low-density Lipoprotein (LDL),* although it is not a significant statistical difference.
- Omega-9 is able to activate *AMP-activated protein Kinase (AMPK),* which happens to be involved in the body's energy balance.
    - *AMPK* triggers fatty acid uptake into the muscle as well as the oxidation of those fatty acids; these fatty acids play a role as one of the body's energy sources.
- *Oleic Acid* can bind itself to 3 *Peroxisome Proliferator-activated Receptors (PPAR). PPAR* is what links fatty acid metabolism and inflammation signals.
    - *PPAR* also regulates oxidative stress, particularly *Reactive Oxygen (ROS)* and *Reactive Nitrogen (RNS) Species. ROS* and *RNS* can cause oxidative damage to organs and tissues and cells.
    - *PPAR* also regulates *Signal Transducer and Activator of Transcription 1 (STAT1); STAT1* promotes an anti-viral environment within select cells. It also has a role in immune response and a pathway that can induce inflammation.

Oleic acid, an omega 9 fatty acid, reacts with the liver's Uncoupling *Protein 2 (UCP), which* can, in turn, provide an antioxidant like effect in the liver and various other organs.

- *UCP* seems to have a direct function controlling the mitochondria's production of *ROS.*

Proposed benefits (evidence varies)

- Antioxidant like effect
- Reduce inflammation
- Increased energy
- May lower cholesterol

Possible Side Effects

There are currently no recorded side effects from regular consumption.

Dosages

The American and European diet already consumes enough Omega-9 fatty acids; supplementation is generally not needed; if one must supplement it, please follow a medical professional's advice.

WARNING

Seek medical advice before use. There is concern for those that over-consume the omega-9 fatty acid erucic acid; erucic *acid* can lead to the condition **thrombocytopenia; this** results in a blood clotting difficulty.

Pantothenic Acid

Pantothenic acid, also known as vitamin B5, is a water-soluble essential nutrient that's widely available throughout the day to day food consumption. Pantothenic acid deficiency is extremely rare and further supplementation does not seem to add any additional benefits from regular consumption; acne is not an established deficiency sign, and routine high-dose treatment lacks sufficient evidence. A variation of pantothenic acid, *dexpanthenol,* is generally more stable and can be found in various cosmetic products.

Pantothenic acid is involved in the production of *Coenzyme A (CoA). CoA* is generally used by most enzymes in the body and is fairly crucial in energy production. *CoA* is the primary out of five coenzymes needed in the *Citric Acid/Krebs Cycle.* The *Citric Acid Cycle* is used to release stored energy through the oxidation of *acetyl-CoA.*

- *CoA* becomes *acetylated* by the breakdown of carbohydrates and fatty acids. Once this is accomplished, *acetyl-CoA* is oxidized to CO2, reducing electron carriers. *Acetyl-CoA* contains a thioester bond whose hydrolysis can release usable free energy*; the energy* released is then used for **ATP** and *GTP (GTP is used for RNA synthesis).*
- The fact that *CoA* is a *thiol* molecule means it can interact with *carboxylic acids. Fatty acids* are considered *carboxylic acids.* This allows *CoA* to transport *fatty acids* to the mitochondria, which are vital for cell structure, or they can also be used for energy purposes.
    - *Carboxylic acids* can create thioesters; this would allow *CoA* to become *acyl-CoA*, an *acetyl-CoA derivative; this* creates the unique ability for *CoA* to carry *fatty acids* into the mitochondria, and if the need arises, oxidize to further release energy.

For those who suffer from acne, Pantothenic acid has been seen to ACTUALLY reduce overall lesion size and redness. This is due to reduced *sebum (oil secretion)* production resulting in smoother skin from decreased solidified *sebum* plugs. This process has been observed to take some time; benefits were observed at 4 months for cheeks and 6 months for the nose and forehead area. It is believed the reason for this benefit is related to the mechanism mentioned above, but a general deficiency in modern diets is not established. Efficient fatty acid metabolism for energy could, in turn, reduce excess lipids in the *sebaceous* glands, thereby reducing a favorable environment for *Propionibacterium acnes (bacteria).*

Benefits

- May reduce acne
- May reduce cholesterol
- General health

Possible Side Effects

- Loose stool
- Hunger pangs
- Dry skin

Dosages

- Adult adequate intake is 5 mg/day. High-dose acne regimens such as 10–12 g/day are not established routine treatment and can cause diarrhea.

WARNING

Seek medical advice before use. Pantothenic acid may affect people with **hemophilia** as it can reduce clotting. Pantothenic acid can also increase the likelihood of **gastrointestinal blockage** if you have a history of **stomach blockage**.

Phenylalanine

Phenylalanine is an essential *aromatic (contains an aromatic ring)* amino acid and a precursor to **tyrosine** and neurotransmitters *dopamine, noradrenaline* and *adrenaline (referred to as catecholamines)* as well as the *neuromodulator* **phenethylamine**. It has been marketed as a nootropic, *analgesic (pain reliever)* and an antidepressant. Supplemental antidepressant or analgesic benefit is not established; phenylalanine supplementation is inappropriate in phenylketonuria.

Phenylalanine is capable of crossing the blood-brain barrier; this is vital as a good portion of *neurotransmitters* in the brain need *phenylalanine* to be synthesized. The importance of this is explained in the following; also, if you have tried researching phenylalanine you probably have found people stating the exact same thing. It is converted to this and converted to that. Let's get into the details and save you some time.

- L-phenylalanine goes through *hydroxylation (an oxidative process)* which results in l-tyrosine; l-tyrosine also goes through *hydroxylation* resulting in **L-DOPA,** which is then *decarboxylated (removal of a carbon atom from a carbon chain)* into *dopamine* and, dependent on the need and cell type, *dopamine* may further be converted into *noradrenaline* or *adrenaline.*

There are three marketed types of Phenylalanine: L-Phenylalanine, D-Phenylalanine, and DL-Phenylalanine.

- D-phenylalanine seems to have the *analgesic* effect mentioned earlier. This may be accomplished as D-phenylalanine may block the body's *nociception (the body's response to harmful situations, i.e., tissue damage/irritation)* response by inhibiting/reducing *enkephalin (regulates nociception)* degradation.
- L-Phenylalanine can promote *Glucagon-like Peptide-1* and *Peptide Tyrosine Tyrosine (PYY) expression; these* peptides are released after food consumption and can reduce *ghrelin* levels, reducing appetite. The *Ghrelin* receptor is also called the *growth-hormone secretagogue receptor, distinct from the growth-hormone receptor,* and plays a role in hunger and body weight.
- DL-Phenylalanine has been seen to contain antidepressant effects which may be due to its interaction with *dopamine* and *serotonin;* however, its mechanism hasn't been fully researched.

Proposed benefits (evidence varies)

- Mood enhancer
- Appetit Suppressant
- Analgesic
- Antidepressant

Possible Side Effects

- Fatigue
- Nausea
- Headache
- Dizziness
- Anxiety
- Constipation
- Heartburn
- Jitteriness
- Sedation

Dosages

- 75-200 mg a day for depression or analgesic effects

WARNING

Seek medical advice before use. Phenylalanine may affect people with **high blood pressure** due to its interaction with *catecholamines.* Those with **neurological conditions** can have adverse side effects due to the same interaction. In those with phenylketonuria, phenylalanine can build up in the body, which can be fatal.

Phenylethylamine

Phenylethylamine is a *neuromodulator alkaloid* and a *trace amine (a family of neuromodulators)* that is synthesized by the amino acid **phenylalanine**. It is marketed and sold as a mood enhancer. Generally, the majority of information that one may find is primarily dedicated to the effects experienced through other methods of supplementation aside from oral consumption, as the majority of phenylethylamine will be broken down in the *First Pass Metabolism* in the small intestine. This may limit oral consumption effects outside of increasing dosages for effect.

Phenylethylamine seems to be synthesized at the same rate as dopamine, although phenylethylamine concentrations in the brain are seen to be three times lower than dopamine due to its fast *MAO-B (Monoamine oxidase B)* enzyme metabolism, with a half-life of about 30 seconds.

- Phenylethylamine can be taken up by *dopaminergic* neurons, which has been seen to increase *dopamine* production while reducing uptake into the neurons; this results in an increase of extracellular concentrations of *dopamine.*
    - This may result in mood enhancement; this being the reason why phenylethylamine is the basis for many drugs. Its carbon main chain/backbone can be modified to make various *substituted phenylethylamines (derivatives of phenylethylamine), the most notable being methamphetamine* and *MDMA (3,4-methylenedioxymethamphetamine).*

Phenylethylamine plays a role in general wellness as it has the capability to regulate mood, attention, and physical energy. Phenylethylamine activates *Trace amine associated receptor 1; this* can inhibit the re-uptake of *dopamine, serotonin,* and *noradrenaline* which may be attributed to that runner's high; amphetamine has been seen to activate the same receptor. It's been noted that *phenylacetic acid (a metabolite of phenylethylamine)* is found at lower than normal concentrations in depressed and bipolar individuals. *Phenylacetic acid has* been seen as a good biomarker that reflects phenylethylamine concentrations.

- *Phenylacetic acid* has been found at higher concentrations in urine after physical exercise, which could contribute to the euphoric/antidepressant effects or "good feeling" that is described after rigorous activity.

Proposed benefits (evidence varies)

- Enhance mood
- Antidepressant
- May enhance exercise
- "Euphoric" feeling
- Increased focus

Possible Side Effects

- Rapid heart rate
- Anxiety
- Agitation

Dosage

- No established safe and effective oral exercise dose exists; user-reported 500–2,000 mg regimens are not clinical dosing standards.

WARNING

Seek medical advice before use. Phenylethylamine may affect people with **dopaminergic disorders** such as **bipolar disorder** or **schizophrenia** as phenylethylamine may increase dopamine secretion resulting in a manic episode. In the same manner, **schizophrenia** symptoms like hallucinations may worsen. Avoid combining with MAO inhibitors; greatly increased exposure can cause serious cardiovascular or neurological reactions.

Phosphatidyl Serine

Phosphatidyl Serine is a phospholipid that's part of cell signaling and has been marketed as a cognitive enhancer, especially in the elderly as it is found in high concentrations in the brain. Phosphatidyl Serine has a similar structure to that of a *triglyceride* except that a phosphatic acid and serine amino acid occupy the third spot where the last fatty acid would be. The majority of research that is out there has been geared towards bovine phosphatidyl serine while the majority, if not all, of phosphatidyl serine that can be purchased is soy-lecithin based. The reason for this is that there was risk of transmitting mad cow disease from the bovine version as it was derived from the brain of bovines. While the phosphatidyl serine is derived from soy-lecithin, it's structurally similar to the bovine one although it's not identical and for practical purposes it may not have beneficial effects at all. It is important to note that most tests were done with bovine extracted phosphatidyl serine.

Phosphatidyl Serine can be found in high concentrations in the brain; once there, it is used in the cell structure as part of the cell's membrane. This structure layering can increase calcium and potassium ion enzyme sensitivity while reducing *acetylcholinesterase (increases the breakdown rate of the neurotransmitter acetylcholine)* receptors. This may promote higher *acetylcholine* levels; this effect may be more prominent in elderly individuals.

- This mechanism of action may be the reason as to how it promotes enhanced cognitive function as it increases ion response sensitivity while also reducing the breakdown of neurotransmitters.

Phosphatidyl Serine seems to have a synergistic effect alongside **fish oil** when taken side by side, as well as assert anti-stress properties. Although the mechanism of action is not yet known as it seems to act independently of cortisol levels and does not affect heart rate; these are textbook biomarkers of stress on the body.

- Phosphatidyl serine can attach itself to two fatty acids being that it's a *phospholipid,* like *docosahexaenoic acid (DHA)* and *eicosapentaenoic acid (EPA),* found in fish oil; this has been seen to increase memory retention; this effect can be seen in a mixture of elderly and young patients alike. Generally, it is preferred that it is bought in this form, with *DHA* and *EPA* in the same bottle as Phosphatidylserine. (Reference **Fish Oil** or **Omega-3** for an in-depth explanation)

Benefits

- Improved cognitive function
- Improve age related cognitive decline
- Improve memory retention

Possible Side Effects

- Insomnia
- Upset Stomach
- Gas

Dosage

- 100 mg 3x a day
- 300 mg with 37.5 mg of **fish oil** for memory retention and improvement in elderly individuals

WARNING

Please consult with a medical professional prior to administering Phosphatidyl Serine. There is concern that Phosphatidyl Serine derived from neural tissue such as bovine tissue can carry prion proteins that can cause the neurodegenerative disease, mad cow disease, also known as **Creutz-Jakob disease**.

Policosanol

Policosanol is made up of the waxes and oils from sugar canes, more so long-chain alcohols; it was marketed as a dietary supplement for lowering cholesterol lipid levels in Cuba. Early favorable findings came largely from Cuban studies; independent replication has been inconsistent. Early Cuban trials reported large lipid reductions, while trials in North America and Germany often found no effect in people with *high cholesterol; these are not comparable success-rate estimates. Independent randomized trials have not consistently reproduced early cholesterol findings; policosanol is not an equivalent substitute for prescribed statins.*

The active component of Policosanol is believed to be *1-octacosanol.* Currently, a synthetic version of *1-octacosanol* has been made and has shown effectivity at inhibiting *HMG-CoA reductase (rate-controlling enzyme of cholesterol).*

- It is believed to be able to stop cholesterol in the same manner that statins do by inhibiting the formation of *mevalonic acid,* which is a precursor to the production of cholesterol in the *mevalonate pathway.* This may be accomplished through inhibition of the enzyme *HMG-CoA reductase* in the same target of *statins (a cholesterol drug).* But generally, this effect seems to be active or augmented through exercise due to liver stimulation.

Currently, policosanol is being studied as an aid to help reduce neuronal damage after a stroke. It has been seen in rats and gerbils to reduce the damage and symptoms of damage after cerebral *Ischemia Reperfusion Injury (interruption of blood and oxygen to the brain).*

- Policosanol has been seen to inhibit platelet aggregation as well as have antioxidant effects.
- Its neuroprotective effects may also be attributed to its previously mentioned ability to reduce/inhibit lipid peroxidation. Data into its mechanism is still preliminary.

The mechanism of its effect isn't fully understood as damage to neural tissue after *ischemia* has various factors. One of them is an excess of neurotransmitters released, which results in metabolic cell death. Another factor is the central nervous system is also involved; this can result in various responses releasing *eicosanoids,* free-radicals and immune responses that can lead to inflammation of the affected area that may amplify cell death or damage.

Proposed benefits (evidence varies)

- Anti-coagulant
- Possible Antioxidant
- Neuroprotective

Possible Side Effects

- Headaches
- Stomach Discomfort
- Skin redness
- Dizziness
- Insomnia
- Weight loss

Dosage

- 5-20 mg a day
- 100-200 mg per 2.2 lbs/1 kg of body weight was used in testing, although this is not recommended for human measuring until human clinical trials have been accomplished.

WARNING

Seek medical advice before use. Policosanol may affect people with **bleeding disorders** as it can increase the risk of bleeding.

Pomegranate *(Punica granatum)*

Pomegranate comes from the *Lythraceae* family, where the ling or bat nut can be found, and can also be found cultivated around the world from the Middle East, Asia all the way to the United States. It is carried by the majority of supermarkets and grocery stores. The edible seeds are called *arils.* Generally speaking, eating the fruit itself has good health benefits, and for the most part, is a better choice of fruits, amongst other choices. In this breakdown, there will be a focus on a couple of its main bioactive compounds, *punicalagins,* that are found in pomegranate juice and *punicic acid* found in pomegranate oil. It is marketed as a superfood due to its antioxidant and anti-inflammatory properties.

Punicalagins are extremely powerful antioxidants, although, through regular oral consumption, an extremely low percentage ever enters the bloodstream. Its metabolites *ellagic acid, gallic acid,* and *urolithins* can still offer antioxidant effects; however, they are not nearly as powerful as *punicalagins.*

- *Punicalagins and their* derivatives seem to promote an anti-inflammatory effect in the intestines via back end activity reduction of *interleukin-6, cyclooxygenase-2* and *prostaglandin E2* instead of front-end expression, while also positively interacting with intestinal *microflora (good bacteria).*
- The extract has been seen to inhibit glucose uptake while providing antioxidant effects against glucose-induced cell stress which can cause *caspase-3 (involved in executing apoptosis/cell death)* and *caspase-8 (involved in apoptosis)* activity; this can provide antidiabetic effects.

Punicic Acid is a fatty acid, that is used and found in pomegranate oil. Interestingly enough it seems *punicic acid* can be converted into **Conjugated Linoleic Acid** in the body.

- *Punicic Acid* is able to reduce *triglyceride* levels in *hyperlipidemic* adults while creating a preferential *high-density lipoprotein (HDL) to triglyceride* profile.

It is important to note that while pomegranate is a healthy fruit, the main bioactive components found in extracts may work for some, while others may see no benefit. This is a direct result from intestinal microflora and varies from person to person as some may not have the bacteria needed to metabolize the bioactive compounds

Benefits

- Antioxidant
- Antidiabetic
- Anti-inflammatory
- Promote preferential cholesterol ratio

Possible Side Effects

There are currently no documented side effects aside from sensitivity/allergic reactions to the extract and may lead to itching, runny nose, difficulty breathing, and/or swelling.

Dosage

- 800 mg of pomegranate extract with 300-320mg of punicalagins may provide antioxidant effects
- 400 mg of pomegranate oil that contains at least 70% punicic acid

WARNING

Please consult with a medical professional prior to consuming Pomegranate; there is concern that Pomegranate may affect those with **plant allergies** as they may be more sensitive or have an allergic reaction to either the fruit or pomegranate extracts. Those diagnosed with **low blood pressure** may be affected by pomegranate extract or the juice as it can lower blood pressure even further.

Potassium

Potassium is a chemical element found on the Periodic Table sitting at atomic number 19. It is in the Alkali Metal group where lithium and sodium are found, and it is a malleable silver-ish white metal and gives off a lilac-colored flame. It is an essential mineral as Sodium and Potassium play hand in hand in total body fluid volume. Sodium works at the *extracellular (outside the cell)* level and potassium works in the *intracellular (inside the cell)* level. It can be seen used as an aid with cardiovascular health due to this reason.

Potassium seems to control blood pressure through its interaction with water fluidity within the cells and the cardiovascular system. Mainly it is believed that increased potassium intake can inversely reduce sodium in the body leading to lower blood pressure. It is also believed that Potassium can regulate cardiovascular sensitivity to *catecholamines (like dopamine, noradrenaline, etc.)* which can affect overall dilation or constriction of blood vessels.

- This may be due to regulating *Transforming Growth Factor-β1 (TGβ1, involved in blood pressure regulation)* expression. High amounts of *TGβ1 have been* implicated in arterial stiffness.

Potassium intake seems to play a role in calcium retention within the bones as lower potassium intake has been implicated in lower calcium retention in the bones.

- Potassium has been seen to increase the expression of *miRNA (micro RNA) miR-223* and *c-miR-21-5p; miR-223* is involved in *osteoclastogenesis (differentiation or better put the process of designating bone cell type). c-miR-21-5p,* on the other hand, has been implicated in protecting *chondrocytes (creates the matrix for cartilage)* from degradation by decreasing *ADAMTS5*, an enzyme implicated in the breakdown of cartilage. This is believed to be the reason for the decreased amount of *u-NTX* and *P1NP*, biomarkers for bone turnover, seen with potassium supplementation.

Potassium may also be linked to reduced kidney stones; this may be due to the body retaining calcium within the bones instead of allowing it to circulate throughout the body to be filtered in the kidneys, thus creating kidney stones. The mechanism above may be the reason for reduced kidney stones.

Benefits

- Promote cardiovascular health
- Reduce the risk of kidney stones
- Promote Mineral Bone Density
- General health

Side Effects

- Upset stomach
- Gastrointestinal discomfort
- Irregular heart beat
- Diarrhea
- Intestinal gas
- Paralysis
- Vomiting
- Confusion

Dosage

- Adult adequate intake: men 3,400 mg/day, women 2,600 mg/day from all sources. These are dietary goals, not supplement prescriptions.

WARNING

Seek medical advice before use. Potassium may affect those with **low blood pressure** due to its blood pressure-lowering effect. Large doses of potassium can also cause irregular heart rhythm and brain damage referred to as hyperkalemia. Kidney disease and medicines such as ACE inhibitors or potassium-sparing diuretics can cause dangerous hyperkalemia; seek individual dosing advice.

Pregnenolone

Pregnenolone, not to be confused with pregnanolone, is a steroid hormone found in the body. It is a precursor to the biosynthesis of most steroid hormones as it shares similar carbon structures. It has also shown biological activity on its own, acting as a neurosteroid; neurosteroids are steroids that are either *endogenous (substances produced within the body)* or *exogenous (substances that are derived outside of the body)* that can quickly affect neuronal activity in the brain.

Pregnenolone can positively regulate *N-methyl-D-Aspartate Receptor (NMDA). NMDA* controls memory and *synaptic plasticity (capability of synapses to strengthen or weaken, the basis of memory and learning.)*

- The sulfate version of pregnenolone, pregnenolone sulfate, can modulate the *glutamatergic* activity of *NMDA*. In short, regulate ion channels that excite synaptic transmission that directly relate to learning and memory.
  - The influx of ions can start a complex chain of events that may activate "silent synapses," synapses that are generally seen as inactive or under-active under regular conditions.

Pregnenolone may also promote an increased release of the neurotransmitter *acetylcholine.*

- *Acetylcholine is* directly involved in memory and muscle control; this action may enhance memory and possibly reduce stress while promoting feelings of relaxation.

Pregnenolone has also been seen to promote *myelination (production of myelin); myelin* is an electrically insulating sheath that surrounds axons of nerve cells.

Pregnenolone has been seen to relieve arthritis symptoms as well as enhance cognitive health; this may be due to the fact that pregnenolone is a precursor to the hormones *glucocorticoids (hormones that can reduce the immune-inflammatory function).* Pregnenolone seems to lower *Low-density Lipoprotein (LDL)* Cholesterol. Reasons as to why are not fully understood at the moment.

<u>*Pregnenolone research/data is lacking, negative effects seen are minimal, interactions are not yet understood*</u>

Benefits

- Enhance cognitive health
- Lower LDL cholesterol
- Promote neuroplasticity
- May aid cognitive dysfunctions
- Promote "positive outlook"

Possible Side Effects

- Restlessness
- Mood change
- Anxiety
- Muscle pain

Dosages

- 25-500 mg a day has been used without any significant side effects

WARNING

Please consult with a medical professional prior to consuming Pregnenolone; there is concern that Pregnenolone may affect those diagnosed with **hormone-sensitive conditions**. Although studies have shown no changes/effects on hormone levels in the body or physical attributes, pregnenolone is a precursor to steroid hormones in the body and may worsen hormone-sensitive conditions.

Protein

When it comes to one of the most important supplements in our modernized world, it must be clearly defined and known. Surprisingly it is not **apple cider vinegar** or **some superfood** but rather something all living matter is made from, proteins. The fact of the matter is that protein needs vary with age, health and training; inadequate intake is possible, but deficiency should not be assumed in the average person. For many in modernized societies, in an attempt to fill those hunger/protein voids, fast foods are chosen for their quick, efficient ease of access that eventually ends up hindering their health further than having had lower protein portions included in their day. Protein bars or shakes are quick and efficient and can provide a wide array of health benefits while curbing hunger until your next REAL meal, all of this while also meeting your daily requirements of protein.

Here are some key points before we get started in the different types and states of protein.

1) During weight loss, higher protein intake can help preserve lean mass and curb hunger. Needs depend on energy deficit, body composition and training; roughly 1.6 –2.4 g/kg/day is used in athletic weight-loss settings.
2) When trying to increase size or mass, 0.5-0.9 grams per lb of bodyweight along with a calorie surplus.
3) For most exercising adults, about 1.4–2.0 grams of protein per kilogram of bodyweight per day supports muscle growth and maintenance.
4) Your body can keep absorbing protein all day after training not just within the first hour of training.
5) **Leucine** helps signal muscle protein synthesis, but 1.8 grams is not a universal ceiling. All essential amino acids are needed. A meal containing about 20–40 grams of high-quality protein is a common guide; the response varies with age, body size, exercise and protein source. Protein absorbed beyond a meal's peak muscle-building response is still used by the body.
6) Muscle remains responsive to protein for at least 24 hours after exercise. Meal timing can help, especially when training fasted, but there is no universal 1–2 hour deadline; daily intake and its distribution matter.
7) Protein sources differ in digestibility and essential amino acid content. Adequate total protein from suitable foods or blends can meet needs economically. Adding BCAAs alone does not supply all essential amino acids or make every protein equivalent, and no single leucine dose guarantees the same response.

States of Protein

The states of protein are based on the amount, filtration and removal of non-protein parts to achieve different purities; they are as follows:

- **Hydrolyzed/Hydrolysate:** Isolates, concentrates, and blends have large peptide molecules that need to be broken by the body before absorption and utilization can take place. Hydrolyzed protein that has been "partially digested" (AKA Hydrolyzed), as unappetizing as it may sound, this can be beneficial as it can increase absorption and delivery to muscle tissue by the body. The process of *hydrolysis* is accomplished through adding enzymes that break down the protein or with the addition of an acid and heat *(similar to your stomach)*. This results in either amino acid peptides or individualized amino acids; additionally, this *hydrolyzed protein* can be absorbed into the body and delivered to the muscle at a faster rate but faster digestion does not consistently establish better recovery than an adequate dose of intact protein. Leucine content depends on the source protein and product.

- **Isolate:** Isolates tend to generally have the highest protein content when compared to other states of protein, pushing anywhere from 80-90% protein content per scoop. This is achieved through a filtration process that removes various non-protein components such as fats which results in very low to insignificant amounts of fats, carbohydrates, and sugars. This makes isolate a great post work out "meal" that enhances recovery time. Isolates generally bear 11% Leucine content.

- **Concentrate:** Concentrates generally bear 65-75% protein content dependent on the manufacturer and process used; the rest is made up of fats and carbohydrates. Although concentrates are not ideal as a post-workout shake, they are great for use throughout the day and as an addition to missed/replacement meals. As mentioned earlier, it can provide carbohydrates and fats needed throughout the day, something Isolates, and Hydrolysates fall short of. Concentrates also tend to be generally wallet-friendly and are the best choice for those looking to increase their protein intake and maintain balanced health as it can aid with fat loss and calorie reduction.

- **Blends:** A Protein Blend is any product that combines various sources of proteins into one product. This can vary anywhere from various whey proteins or protein types, i.e., Isolate, concentrate, and hydrolysate. Or it can refer to a combination of different types of proteins like egg, milk, whey, and casein protein. The term is used loosely, and there is no direct regulation of proper use. The best bet would be to read the label and compare the ingredients to the types referenced above. The first ingredient is always the most prominent product it is composed of; then the second would be the second most prominent "ingredient" found in the product, so on and so forth.

Here's a quick tip prior to going to the different types of protein: if you want to find roughly the percentage of protein that a powder has, divide the protein content by the serving size. For example, a nutrition label may read, Serving Size 1 Scoop (30g); as you look down the Nutrition Label will show **Protein** 22g and then some percentage like 44%. The percentage is not important. **To find the protein purity/percentage in your powder, divide 22g/30g which is .73333. Move the decimal over to the right 2 times, and you have your percentage**. This protein scoop is about 73% protein, and the rest would be the sugars, carbohydrates, and fats. All proteins have a Biological Value (BV); the higher the value, the more the body can absorb and utilize the protein.

Beef Protein (BV 80)

- Beef protein is a challenging topic in any sports and fitness community, and it's for a good cause. The general concept of protein derived from beef is an ideal product that could potentially benefit many athletes and gym-goers alike. Unfortunately, the biggest downfall in Beef Protein is the lack of beef in the protein powder. Generally seen, most companies do not sell a high-quality beef protein; instead, they sell a **collagen/**gelatin-based powder that does not provide the essential amino acids and has a poor amino acid profile in general. However, some studies have shown that a product with high-quality beef protein has resulted in effective recovery and improvement in comparison to whey. This is not to say that beef protein powder will not provide any benefit; at the end of the day there are still amino acids and proteins in the powder. Pure beef protein can provide benefits comparable to egg, or whey protein; however, poor beef protein falls short of other protein powders, although it is a good source of collagen.

Protein

Casein Protein (BV 91)

- Casein protein, much like whey, is also derived from milk and is a byproduct of cheese making but unlike whey, it takes a longer time to absorb and digest. This provides a sustained amino acid delivery throughout long periods of time, promoting an anabolic environment in the body. This protein is beneficial right before bed or during dieting as it promotes feelings of fullness and curbs hunger. The most common forms of casein found are *milk, calcium caseinate, and micellar casein.* Micellar casein is separated from milk and whey through filtration, and it remains unaltered making it much slower digesting compared to calcium caseinate that is treated with alkaline substances, like calcium to separate it from the milk and whey, thereby altering the protein making it more digestible.

Egg Protein (BV 100)

- Egg protein comes in second compared to Whey Protein in terms of absorption and utilization. It has all 9 essential amino acids and can be easily digested, thereby enhancing absorption and utilization by the body. Egg protein can be found as egg white isolate, or egg protein which contains egg white and yolk; the added benefit to this is that you also receive the health benefits of whole eggs without the high amounts of cholesterol; some brands are up to 98% less cholesterol than regular eggs. Egg yolks have important nutrients such as **choline** which can support cellular and brain function/health. Another added benefit is that it has no lactose for those that are lactose intolerant and is almost on par as effective as whey.

Hemp Protein (BV 87? Not enough reliable data at the moment)

- Hemp protein is a popular choice for protein amongst Vegans and Vegetarians; unfortunately, hemp protein on its own is not a suitable choice for a complete protein replacement as it lacks formidable levels of the 9 essential amino acids, one of them being Leucine. Low levels of leucine can affect protein synthesis as it is one of the main activators of protein synthesis. It does have a decent absorption rate, however, and it is recommended to either take higher amounts to reach biologically active amino acid levels or as a mixture of other plant proteins such as pea or rice protein.

Pea Protein (BV 65)

- Pea protein powders can effectively push 90% protein content; this is easily achieved as the majority of pea constituents are carbohydrates. Once those are removed through filtration and refinement the majority of what's left is protein. Unfortunately, like hemp protein, it lacks biologically active amino acid levels, particularly **methionine**. Pea protein, in combination with rice protein, can provide an effective amino acid profile in a similarly effective fashion to that of whey at a standard of 45% Pea Protein and 55% Rice Protein. This combination provides a complete amino acid profile that can support Vegan and Vegetarian lifestyles.

Rice Protein (BV 83)

- While most perceive rice to be a great choice of carbs to complement a protein in a dish like chicken, rice has been found to make a good source of protein when refined and filtered. Like pea protein, once carbohydrates have been removed from the rice, one of the major constituents, you are left with a protein powder that bears about 70% protein content. Unfortunately, rice protein alone is not a good source of the 9 essential amino acids as it lacks **lysine but is high** in **methionine,** making it complementary to pea protein. The combination of Pea Protein and Rice Protein has a comparable effectivity and amino acid profile to that of Whey Protein; this is able to support Vegan and Vegetarian lifestyles.

Soy Protein (BV 74)

- Soy protein can offer benefits to males and females, and controlled-trial meta-analysis does not show that soy or isoflavones lower testosterone or raise *estrogen* levels in men. Isoflavones can interact with estrogen receptors, but this mechanism does not establish feminizing effects from ordinary soy intake. Soy can be a useful major protein source for Vegan and Vegetarian lifestyles. Like other plants, it contains compounds sometimes called *anti-nutrients.* Some can reduce nutrient absorption, but processing and preparation reduce their effects; adequate varied diets can meet requirements. On the positive side, soy protein does have the proper 9 essential amino acid profile in useful proportions; other plant foods also contain these amino acids in differing amounts making it a complete protein. Soy has mixed anecdotal and scientific reviews; some swear by it, others curse it as heresy. In truth, most of the "effects" that are experienced are just possible side effects that can occur like any other supplement, especially at larger doses. Additionally, trial quality varies, and menopausal symptom findings depend on formulation and outcome. At times, estrogen levels have even been seen to lower with the addition of soy protein. It's good to have a balanced diet and not to solely rely on one protein source, thereby reaping all health benefits from all products.

Whey Protein (BV 104)

- Whey protein has one of the highest Biological Values (BV), meaning it has one of the highest affinities of absorption and utilization by the body. It is easily digested and is a complete protein having all 9 essential amino acids. It is derived from milk, originally as a byproduct of cheese making. Whey protein has been seen to aid in fat loss while enhancing overall health as it can improve **glutathione** (most prominent antioxidant in the body) levels in the body since it's made of L-glutamic acid, L-cysteine and glycine.

WARNING

Seek medical advice before use. There is concern that…

Beef Protein

Beef Protein has not been seen to have any side effects as of current anecdotal and clinical data.

Casein Protein

Casein Protein may affect those that are lactose intolerant, causing side effects such as cramps, bloating, nausea, and bowel movements.

Egg Protein

Egg Protein may affect those with **gastrointestinal problems** causing side effects such as cramps, bloating, nausea, gas, and bowel movements. If you are **allergic** to egg proteins (albumin proteins), egg protein can cause an allergic reaction.

Protein

Hemp Protein

Hemp Protein may affect those with **gastrointestinal problems** as some hemp protein powders can contain high amounts of fiber, causing side effects such as cramps, bloating, nausea, and bowel movements.

Pea Protein

Pea Protein may affect those with **Irritable Bowel Syndrome** as Pea Protein contains small amounts of *Fermentable Saccharides* and *Polyols*; these carbohydrates have been known to worsen symptoms.

Rice Protein

Rice Protein has not been seen to have any side effects as of current anecdotal and clinical data.

Soy Protein

Soy Protein may affect those with **Hormone Sensitive Conditions** due to its *Phyto-estrogen* compounds which can worsen symptoms on diseases such as but not limited to **Endometriosis** and **Breast Cancer**. Soy Protein can also cause side effects such as cramps, bloating, nausea, and bowel movements. Soy Protein may also cause an **allergic reaction** in those with **Hay Fever.**

Whey Protein

Whey Protein may affect those that are **lactose intolerant,** causing side effects such as cramps, bloating, nausea, and bowel movements.

Pycnogenol® (Pinebark extract)

Pycnogenol is an extract derived from the French Maritime Pine *(Pinus pinaster)* bark. Pycnogenol has a high concentration of *procyanidin compounds; these* compounds provide the extract with its bioactivity. The compound itself does not contain the bioactive components; rather, it is converted into the bioactive *procyanidin* compound by bacteria found in the colon. It's marketed as an anti-inflammatory, an antioxidant as well as having possible antidiabetic properties. It is thought that many of its effects increase as it builds up in the body. A Cochrane review found insufficient reliable evidence for chronic disorders; trials were small and heterogeneous, so broad clinical benefits remain uncertain.

Its interaction as an antioxidant can provide a multitude of benefits. Through its antioxidant fashion, it is able to increase *Nitric Oxide* levels in the body by prolonging *Nitric Oxide's* half-life while reducing its conversion into *Superoxide.* It has also been seen to promote *the* production of *Nitric Oxide* through *Nitric Oxide Synthase (enzyme involved in Nitric Oxide production)* interaction.

- *Superoxide Radicals* are toxic and can cause oxidative damage in the body. Generally speaking, Pycnogenol is also able to act as most antioxidants and sequester free radicals in the body such as *Superoxide* and *Hydroxyl* radicals.

Interestingly enough in states of *Nitric Oxide* overproduction, the *procyanidin* metabolites can reduce *Nitric Oxide* overproduction by inhibiting nitrite; this can result in an anti-inflammatory effect.

- These effects have been seen as an effective *vasodilator,* increasing blood flow independent of blood pressure. Additionally, the metabolites can reduce *Interleukin-8 (inflammatory messenger)* in *chondrocytes (cells found in cartilage).*

The antidiabetic effect shows that the extract can increase glucose uptake into fat cells. Although this is seen as an antidiabetic effect, this may also cause obesity. It is able to accomplish this through the *PI3K/AKT Pathway.*

- *PI3K (Phosphoinositide 3-kinase)* is responsible for liver insulin sensitivity and the distribution of glucose to cell functions or particular cells. When PI3K phosphorylates, it activates *PKB.*
- *PKB (Protein kinase B),* also known as AKT, induces glucose transport.

Proposed benefits (evidence varies)

- Aid with osteoarthritis
- Aid with erectile dysfunction
- Anti-inflammatory
- Vasodilator
- Antidiabetic
- Antioxidant

Possible Side Effects

- Dizziness
- Gastrointestinal Discomfort
- Headache

Dosage

- 50-65 mg 3x a day, for osteoarthritis benefits and symptom reduction is not seen until 90 days.

WARNING

Seek medical advice before use. Pycnogenol may affect people with **auto-immune diseases** such as **lupus** and **rheumatoid arthritis** as Pycnogenol does interact with the immune system. Bleeding may be increased in those with **bleeding problems**. **Diabetics** should monitor blood sugar levels as Pycnogenol can affect insulin sensitivity and glucose levels; insulin medication may need to be adjusted accordingly.

Pygeum *(Prunus africana)*

Pygeum is an extract that is derived from the bark of the *Prunus africana* tree belonging to the *Rosaceae* family where apples and cherries can be found. It was generally seen as traditional African medicine and consumed as a tea in an attempt to treat urinary problems in men. Currently, it is marketed and used for prostate health problems such as *Benign Prostatic Hyperplasia (BPH). BPH* is a common condition seen in men over 50, in which the prostate becomes enlarged and blocks the urinary flow; this can cause kidney and bladder problems if left untreated.

The active compound found in Pygeum is believed to be *N-butylbenzenesulfonamide; it* has been seen to act as an *Androgen Receptor (AR) antagonist (in short it blocks and/or stops androgen hormones like testosterone).* Its main function seems to come from its ability to inhibit *fibroblasts* and *myofibroblasts (involved in cell type change) proliferation (cell multiplication)* in the prostate.

- *Fibroblast* stimulation is the process in which *fibrocytes* make the matrix or "space" that's outside of the cell to enhance structural and biochemical support for the cells around it.

The process mentioned above is related to prostate growth, although it may not reverse *BPH,* it seems to be able to inhibit further growth. This same concept may apply to *androgen*-sensitive prostate cancers.

- It is important to note the mechanism of this effect is through *AR* inhibition. This means it will affect *IGF-1* and *Dihydrotestosterone (DHT)* production, which can result in an *anti-anabolic* environment for athletes.
    - *DHT stimulates prostate cell proliferation, converted from testosterone in the prostate gland, leading to an enlarged either benign or malignant prostate. DHT has also been implicated in pattern hair loss.*
- The secondary benefit that is seen is related to its anti-inflammatory activity, as it is able to inhibit *interleukin-6 (IL-6). IL-6* can be a pro- or anti- inflammatory messenger dependent on its activation. Pygeum extract has been seen to inhibit its pro-inflammatory response signaling.

Pygeum also contains *atraric acid. Atraric acid* has been seen to displace *androgens* as it can bind to *AR* and effectively compete with *DHT* and various other androgens. At high concentrations *atraric acid* can become an agonist (activator) of estrogen receptors *ER-α* and *ER-β*; however, at normal treatment doses, it does not seem to affect either.

Benefits

- Anti-inflammatory
- Possible Phytoestrogen
- Possible Benign Prostatic Hyperplasia treatment

Possible Side Effects

- Gastrointestinal discomfort
- Diarrhea
- Nausea

Dosage

- 50-100 mg a day

WARNING

Please consult with a medical professional prior to consuming Pygeum; there is concern that Pygeum may affect those diagnosed with **hormone sensitive conditions** such as **endometriosis**, **uterine fibroids**, **uterine**, **ovarian, breast**, **prostate,** and **testicular cancer**. More data and research need to be done as trials and clinical data are lacking.

Quercetin

Quercetin is a *bioflavonoid* that is found in various fruits and vegetables like tomatoes and apples. Like most *flavonoids*, quercetin is considered an antioxidant that possesses anti-inflammatory, anti-hypertensive, vasodilator effects, anti-hypercholesterolemic and anti-atherosclerotic activities. Generally seen, absorption into the body is moderate at best and varies greatly depending on numerous variables, although consumption alongside *long-fatty chain acids* can increase bioavailability due to its *lipophilic (fat-loving)* properties.

Quercetin can interact with adenosine signaling in experimental models, but human evidence does not establish a practical stimulant effect stronger than caffeine. P*oor* oral bioavailability alone does not explain away the absence of demonstrated clinical benefit.

- Adenosine can create a sedative-like effect by interacting with *adenosine receptors*, and it has been used for similar purposes in hospitals. Blocking the receptors off can cause the exact opposite effect and produce a stimulant effect.

Quercetin can act as a neuro anti-inflammatory by promoting *heme-oxygenase-1* production, which can directly inhibit nitric oxide as well as other inflammatory markers that are directly released in response to inflammation.

- These are preclinical hypotheses; clinical treatment of Alzheimer's or Parkinson's disease is not established. The downside is that quercetin intake may also directly affect long term memory production as it seems it interferes with short term to long term memory conversion.

The benefits stated in the first paragraph are generally achieved from quercetin's effects as an antioxidant as it has very similar protective effects like other polyphenols/flavonoids. Mainly, protection and prevention of oxidation on compounds such as *Reactive Oxygen Species (ROS is a reactive Oxygen that can cause free radicals and damage to the body and cells).* It is also able to inhibit fat accumulation, platelet aggregation improving *endothelium (blood cell lining)* health.

It is important to note that although quercetin is looked upon as a powerful antioxidant, it can also become toxic when above-normal levels are reached within the body.

Proposed benefits (evidence varies)

- Cardiovascular health aid
- Antioxidant
- Vasodilator
- May Aid neuro-degeneration

Possible Side Effects

- Tingling of extremities
- Nausea
- Headache
- Flushing
- Vomiting
- Sweating

Dosages

- 12.5-25 mg per kg/2.2 lbs per body weight

WARNING

Seek medical advice before use. Quercetin may affect people with **Kidney problems**, as quercetin can cause kidney damage at high doses. Quercetin can reduce drug metabolization in the liver as it has interaction with *Cytochrome P450; refer* to a doctor if you are on any medication; this can affect those with **liver problems.**

R-ALA *(R-Alpha Lipoic Acid)*

If you referenced **Alpha Lipoic Acid** earlier in the book, this page will essentially have the same information outside of the fact that **R-ALA** is an isomer of ALA, particularly an *enantiomer.*

- An *enantiomer* is the exact reflection of the chemical structure that is reversed in an axis. Very similar to someone's left and right hand or feet. They cannot be flipped to reach the desired effect of the other, and many times one isomer's effect can be the exact opposite of the other.

The stereoisomers are R-ALA and S-ALA. R-ALA is the naturally occurring form used by mitochondrial enzymes; Na-R-ALA means sodium R-lipoate, not S-ALA. Synthetic ALA often contains both forms. This biochemical role does not establish that every supplemental form improves health or that the R form has superior clinical outcomes for every indication.

- ALA can be seen as a potent *chelator (binds to metals so they can be removed out of the body)* only when metal levels are above normal, such as **iron** and **manganese**.
- It is also able to enhance oxygen consumption, thereby reducing oxygen radicals that cause oxidative damage. At the same time, since it's also crucial for aerobic metabolism (a process of creating energy through oxygen), it is able to help cells convert energy from nutrients into **Adenosine Triphosphate (ATP); this, in turn,** can aid athletic performance and recovery.
- Similar to **NAC**, ALA has the ability to support **glutathione** concentrations, especially in neuronal tissue, having a secondary effect as a powerful antioxidant.
    - Glutathione is an antioxidant responsible for reducing oxidative damage.
- It may be able to reduce appetite while increasing energy levels by increasing *glucose* uptake into the *hypothalamus*. The result of this seems to be brought on by mimicking a calorie surplus through *AMPK (Plays a role in glucose uptake when cellular energy is low)* inhibition in the *hypothalamus*. This can result in weight loss from a calorie deficit diet.
- ALA can also promote cardiovascular health as it can inhibit platelet aggregation (formation/cluster) through various pathways, including **arachidonic acid**-induced aggregation.

It seems to also have a potent anti-inflammatory action, as it is able to inhibit *NF-kB (NF-kB controls genes involved in inflammation)* inflammatory signaling activation.

Proposed benefits (evidence varies)

- Cardiovascular health aid
- Improve cognitive function
- Appetite suppressant
- Chelation
- Weight loss
- Improve energy levels
- Antioxidant
- Anti-inflammatory

Possible Side Effects

- Rash
- Nausea

Dosage

- 300-600 mg a day, **Medium-chain Triglycerides (MCT)** can reduce absorption.

WARNING

Seek medical advice before use. R-ALA may affect people with Diabetes; ALA can lower blood sugar levels. In people with Thyroid problems, ALA might interfere with Thyroid treatments. ALA has been known to lower thiamine levels in the body in a similar fashion as alcohol consumption; taking ALA while **consuming alcohol** can lead to serious health conditions and dangerously low levels of thiamine in the body.

Raspberry Ketones

Raspberry ketones are naturally occurring *phenolic compounds* found in raspberries that are responsible for the raspberry's smell. The *phenolic compounds* can also be found in other berries too such as cranberries and blackberries; they are just more prominent in raspberries. It has been marketed as a fat loss/burning compound as it has compounds similar to *ephedrine* and **synephrine**. Both of these compounds are capable and used for fat loss; similarity does not establish comparable effects or safety from raspberry ketones in humans. The problem for the effects to be prominent from oral consumption alone is reaching high enough concentrations in the body, particularly on fat cells. Theoretically it is possible, although, there currently is no human evidence to prove this. Anecdotes and animal mechanisms do not establish an effective human regimen. Experimental high doses should not be converted into self-treatment advice.

Raspberry ketones seem to be able to activate genes that are involved in *lipolysis (The breakdown of lipids particularly adipose/fatty tissue),* like *adipose triglycerides lipase (ATGL)* and *hormone-sensitive lipase (HSL).*

- *ATGL* has a biological function of lipid storage, metabolism, catabolic process, and organization.
- *HSL* has a biological function of lipid metabolic and catabolic processes along with *lipase activity (enzymes that can cause/accelerate the breakdown of fats).*

When combined with *adrenergic* substances such as synephrine and ephedrine, it may be able to augment the *lipolysis* function of *adipose (fat)* tissue. As it was seen to augment and stimulate lipolysis with *norepinephrine (A neurotransmitter).*

- *It is also important to note that at high doses raspberry ketones can inhibit testosterone production as it can possibly inhibit androgen receptors.*

It may also be able to reduce and suppress the creation/transformation of *adipocytes (fat cells)* by decreasing *adipogenesis (creation of fat cells); it* is also able to accomplish this by reducing the synthesis of fat cells through *gene expression.*

Outside of raspberry ketone isolation, the entire fruit/plant does hold benefits. Raspberry leaf tea has traditional pregnancy uses, but reliable evidence for labor, nausea or lactation benefits and safety is limited; it is not equivalent to isolated ketones.

Proposed benefits (evidence varies)

- Possible weight-loss
- Leaf and fruit differ from ketones

Possible Side Effects

Through a lack of data, these are the only reported side effects: jitteriness, increased blood pressure and rapid heart-beat. There currently is no data for "active dosage" although cell toxicity may be a concern.

Dosage

- The 100–340 mg/kg amounts come from non-human experimental work, not a recommended human regimen. Human weight-loss efficacy and safety are inadequately established.

WARNING

Seek medical advice before use. No established effective or safe weight-loss dose exists. Do not try to reproduce animal exposures. Raspberry leaf, whole fruit and isolated ketones differ; pregnancy use of one cannot be justified by claims for another.

Red Yeast Rice *(Monacolin K)*

Red yeast rice is basically as it sounds, it's rice that is red with yeast. Its origins are mainly seen in traditional Chinese Medicine. If you were to hold the rice or the paste in your hand it does have a reddish color to it with a purple-ish tint to it; the color itself comes from being fermented with the mold, *Monascus purpureus.* It is claimed to be a potent natural statin comparable to *Lovastatin.*

- *Statins* are medications used to reduce the risk of cardiovascular diseases and their mortalities, by lowering *lipids* such as cholesterol.

The active component found in Red Yeast Rice is *Monacolin K.* This compound can be isolated from the rice after it has been fermented. When isolated, it has an identical chemical structure comparable to that of the prescription drug *lovastatin.* *FDA states that products with enhanced or added lovastatin cannot be marketed as dietary supplements; legality does not depend simply on naming Monacolin K on the label.* The basic mechanism of it is the same as any other statin; it can lower *lipids* through inhibiting *HMG-CoA Reductase.*

- *HMG CoA reductase* is an enzyme that produces cholesterol by the formation of *mevalonic acid,* which is a precursor to the production of cholesterol in the *mevalonate pathway.* *Monacolin K's benefit* has been seen to inhibit *HMG-CoA reductase* like other statins; this effect seems to be augmented through exercise due to liver stimulation. Inhibition of this enzyme increases translocation of cholesterol into cholesterol sensitive membranes and lowers levels of cholesterol in the body. Cholesterol translocation has been seen to be consistent, however mildly effective.

Commercial products vary widely in *Monacolin K* content, often without declaring the amount; an unlisted amount is not a guarantee of legality, efficacy or safety. The only concern with this is the unknown amount that one would be consuming. Another concern is the *Citrinin* content. *Citrinin* can be found in some Red Yeast Rice products and is a toxin that is produced by some fungus that is capable of causing death or severe illnesses in humans. The kidney toxicity has also been observed as these particular metabolites are able to target some renal cells, which could eventually lead to kidney failure. *Citrinin* is also able to cause gene damage in cells, particularly *lymphocytes (white blood cells).* Gene damage in the cells can cause cell mutation which could ultimately lead to cancer. Another method of intake is consumed as food. Red Yeast rice is used for many dishes such as pickling, pastries and even some wines. Food use does not establish a predictable therapeutic dose or remove statin-like or contamination risks.

Proposed benefits (evidence varies)

- Support Cardiovascular Health
- Reduce Blood Pressure
- Natural statin

Possible Side Effects

- Headache
- Stomachache
- Heartburn
- Muscle aches
- Dizziness

Dosages

- No universal safe supplement regimen is established. Rice weight is not Monacolin K dose; discuss cholesterol treatment and statin-like interactions with a clinician.

WARNING

Seek medical advice before use. Red Yeast Rice may affect those with **Liver Problems**, as Red Yeast Rice reduces cholesterol synthesis in the liver. Be aware of the toxic metabolites as well as explained above in the underlined section.

Resveratrol *(3,5,4'- trihydroxystilbene)*

Resveratrol is a *natural phenolic* compound produced on the skin of various berries to include grapes and raspberries particularly when it's under attack by fungus and bacteria, as resveratrol is an anti-fungal, anti-microbial and an anti-oxidative. There are two isomers of resveratrol: one is *trans-resveratrol* and the other is cis-resveratrol; both are antioxidants, while *trans-resveratrol* seems to be the more potent of the two. Natural *phenols* are made by plants in response to ecological stress to protect the plant from its environment.

Resveratrol has low oral bioavailability because of rapid metabolism, and while it does have a plethora of benefits, beneficial concentrations may be hard to achieve. Preclinical studies explore effects on cancer cells and disease pathways, but cancer prevention or treatment in humans is unestablished as it can influence *Nuclear Factor kB* (*NF-kB)* and *PI3K/Akt/mTOR* pathways and signaling. *NF-kB controls* DNA transcription, immune function, and cell survival.

- *NF-kB* dysregulation can support cancer-cell survival and chronic inflammation, but is not a single primary cause of cancer or a direct mutat*ion mechanism.* Additionally, *NF-kB* controls immune cell signaling through *cytokine (immune cell/messengers)* production. Improper or over *cytokine* production can lead to inflammatory and *autoimmune* diseases. This results from over-inflammation or an over-reactive response by white blood cells which would, in turn, start attacking the body.
- *PI3K/Akt/mTOR* is the intra-cell signaling, or signaling within the inside of the cells. It plays a key role in regulating a cell's cycle, which involves division and duplication of DNA. In many cases this pathway/signaling is seen to be overactive in cancers; this leads to the *proliferation (cell multiplication)* of cancer cells, while also reducing *apoptosis (programmed cell death)* of those cells.

These mechanisms can have a snowball effect on each other and affect various *neurodegenerative* and *metabolic diseases*, as well as cancers and cardiovascular disorders. Since resveratrol can positively regulate and influence these pathways, the mechanisms are research leads rather than proof of a wide range of clinical benefits.

Proposed benefits (evidence varies)

- Cardiovascular health aid
- Promote neuronal health
- Anti-cancerous
- Anti-fungal
- Antioxidant
- Anti-inflammatory

Possible Side Effects

Currently, there is a lack of data and anecdotal reported side effects; no current side effects are currently known.

Dosages

- 5-10 mg a day for general health benefits and long-term health

WARNING

Seek medical advice before use. Review bleeding disorders and anticoagulants because resveratrol may affect clotting. Experimental phytoestrogen activity warrants advice with hormone-sensitive conditions, including **uterine, ovarian or breast cancer**, **endometriosis** and **uterine fibroids**.

Rhodiola rosea

Rhodiola rosea, also known as Rosenroot, Golden Root and Arctic Root, is a flowering plant from the *Crassulaceae* family where the Succulent plants (plants with thick, fleshy leaves) can be found. It is native to Arctic environments on any continent. It has been used in Scandinavian Traditional medicine for physical adaptation to harsh climates and environments as well as for mental stress and fatigue; as such, it falls under the category of *adaptogen*. Viking use is claimed; human clinical and performance benefits remain unestablished.

As of current research and data, there isn't a clear active component for the stated benefits, although a few do stand out: salidroside, *rosavin*, and *tyrosol*. *In some labels, you may see the name rhodioloside or rhodosin, both of these are the same as salidroside. You more than likely will also see* ***SHR-5 listed; this*** *is a standardized extract that has been successfully used in human studies.*

- The extract can inhibit *Monoamine oxidase A (MAO-A)*. This enzyme is responsible for normal brain function as it is able to degrade and oxidize amine neurotransmitters such as *dopamine, serotonin,* and *norepinephrine*. Elevated levels of *MAO-A* can lead to increased stress and depression, as well as other cognitive problems and disorders.
- Additionally, it has been seen to enhance *neuropeptide y* activity in the body, which is considered an *anxiolytic (anti-anxiety)* peptide. It may also be seen as a stress modulator.

A combination of both activities, *MAO-A* and *neuropeptide y*, may be the reason for its anti-stress properties. Its cognitive enhancing abilities may be derived from the reduction in oxidative stress in neural stem cells; this, in turn, can promote normal *neurogenesis (nerve cell creation)* rates. This leads to an antioxidant effect in neurons, as it is also able to promote *mitochondrial* function which in turn can enhance oxygen consumption leading to a reduction in *Reactive Oxygen Species (ROS)*. *ROS* are oxygen radicals that can cause oxidative damage to nearby cells.

- This antioxidant effect seems to be carried over throughout the body as it offers cardiac tissue protection as it has been seen to protect *cardiomyocytes* from *mitochondrial* induced death derived from *ischemia (oxygen/nutrient restriction to cells)*.

Proposed benefits (evidence varies)

- Reduce mental fatigue
- Reduce physical fatigue
- Aid in work-related stress
- Antioxidant
- Post-workout recovery
- Cognitive enhancer
- Anti-anxiolytic

Possible Side Effects

- Dry mouth
- Saliva over-production
- Dizziness
- Dizziness

Dosage

- 280-680 mg of SHR-5 after an exhaustive and/or prolonged stressful event

WARNING

Seek medical advice before use. Review **autoimmune disease and diabetes** medicines. Salidroside has experimental glucose-lowering effects; its glucoside structure does not establish blocked sugar absorption. Monitor glucose without self-adjusting treatment.

Ribose

Ribose is a carbohydrate, specifically a monosaccharide, the most basic of sugars. Generally, most people think carbohydrates are complex sugars, but they can also be basic and simple sugars like *monosaccharides* and *disaccharides.* Ribose plays a role in energy production and has been marketed to enhance **Adenosine Triphosphate (ATP)** production and recovery since it can enhance ATP recovery in the body. Ribose additionally makes up the backbone of *RNA (Ribonucleic acid),* while *deoxyribose, with one fewer oxygen atom,* forms the sugar backbone of *DNA (Deoxyribonucleic acid).* A *saccharide* is just a sugar with the prefix stating how many are attached.

It's important to note that human performance findings are inconsistent in both trained and untrained people; a structural role in ATP does not establish supplemental benefit. It is also important to note that there have also been mixed results in the following benefits. Ribose intake is fairly low in the American diet, and supplementation can lead to an enhanced recovery in new/untrained individuals.

The belief behind enhanced ATP recovery is the fact that ribose is part of ATP's molecular make-up, as it attaches *adenine* to the *triphosphate* group.

- In theory, as ATP is broken down and used for energy in high-intensity events and training, it must be replenished through various enzymes and phosphate pathways. One of these enzymes, *glucose 6-phosphate dehydrogenase (G6PDH),* limits the rate at which new ATP is created. Ribose is believed to bypass this enzyme altogether as it is used by the *Pentose Phosphate Pathway; this* pathway can create *ribose-5-phosphate (used for nucleic acid synthesis)* and *5-phosphyl-ribose-1-pyrophosphate* which can be used in the synthesis of ATP thereby shortening ATP recovery time.

Additionally, when ATP is broken down into *inosine 5'-monophosphate (IMP),* it becomes basically useless as some will also leave the cell. Some of the remains can be salvaged and reused to create ATP. The secondary way to replenish ATP is through new ATP synthesis, which is extremely slow and can take up to a few days to complete. When ribose is introduced into the muscle, it is able to enhance *nucleotide synthesis* as it needs a 5-carbon sugar which is exactly what ribose is. ATP is also a nucleotide that is created through this very process.

Proposed benefits (evidence varies)

- Improved recovery time
- Increased endurance
- Reduced muscle fatigue

Possible Side Effects

- Gastrointestinal discomfort
- Nausea
- Diarrhea
- Headache

Dosage

- 3-5g a day, as it builds up in the body recovery should become more apparent

WARNING

Seek medical advice before use. Ribose may affect **diabetics** as ribose is a sugar and may affect blood sugar levels; consult with a doctor to modify medications.

Rose Hip (Dog Rose)

Rose Hip generally refers to the fruit found in the Rose plant that can either be used as a standalone product for teas, jellies and such or extracted and used in a range of supplements. Although there are various species of roses that can be used, the most well-known for its benefits is *rosa canina*. As such most supplements are derived from the *rosa canina* rose, also referred to as the Dog Rose, which is a wild rose growing in parts of Europe, Northern regions of Africa and the Western Regions of Asia. It has been claimed to relieve various symptoms in *rheumatoid* and *osteoarthritis.*

Interestingly enough, the compounds found in the extract synergistically enhance each other's properties. The only active compound that may be special to rose hip is Galactolipids, which only seem to be beneficial when included with the entire compound matrix found in rose hip. Its main mechanism has been seen to relieve these symptoms through immune suppression of the inflammatory *cytokine (messenger signal) interleukin 1-Beta (1L-1β).*

- *1L-1β* can directly stimulate *matrix metalloproteinase (MMP) -1, -3 and -13* release. These particular proteins are responsible for *Type-II* ***Collagen*** turnover. *Type-II Collagen* is collagen that is found and used to create the cartilage found on the surface of joints. An overactive release of these proteins is what can lead to joint degradation found in arthritis.
- *1L-1β* can also stimulate the release of *Leukemia Inhibitor Factor (LIF) protein; LIF* can be found on joints; over-expressed LIF, however, can lead to the breakdown of cartilage and increase the *pathogenesis (development of a disease)* of arthritis.

Granulocytes (a type of white blood cell), particularly *neutrophils,* can migrate to joints and make up to 90% of the cells found in the synovial fluid in those with rheumatoid arthritis. *Neutrophils* are responsible for *lysosomal hydrolases* release; *lysosomal hydrolases* can cause injury and oxidative damage to the surrounding tissue. Rose hip has been seen to suppress the migration of *granulocytes* towards synovial fluid, as *neutrophils* are attracted to inflammation, as they are one of the first responders to stimuli.

Benefits

- Rheumatoid arthritis aid
- Osteoarthritis aid
- Antioxidant
- Anti-inflammatory

Possible Side Effects

- Nausea
- Vomiting
- Stomach cramps
- Diarrhea
- Constipation
- Heartburn
- Fatigue
- Headaches

Dosage

- 2-5g twice a day

WARNING

Please consult with a medical professional prior to consuming Rose Hip; there is concern that Rose Hip may affect those with Vitamin C complications such as **Sickle Cell** as Rose Hip has a high Vitamin C content and can cause a Sickle Cell Crisis. Vitamin C can also affect any **Iron Deficiency/Disorder,** as Vitamin C can also increase iron absorption. Rose Hip also contains Rugosin E., which can increase the risk of clotting especially those with a clotting history such as **stroke, heart attack,** and **blood clots**.

SAMe *(S-Adenosyl methionine)*

SAMe is made up of **methionine** attached to an **ATP** molecule and can be found and used all throughout the body. Its main function is to donate a methyl group to other molecules. This action provides support to all *metabolic* functions in the body. *Metabolic* functions can be thought of as: food to fuel conversion for energy, elimination of wastes, repair, and rebuild of numerous tissues, etc. The supplementation of SAMe has been marketed as an aid for people with osteoarthritis and depression. Clinical evidence for depression, osteoarthritis and liver disease remains inconclusive; methyl donation alone does not establish a treatment effect.

Low levels of SAMe in the body can lead to poor maintenance of joints and connective tissues as well as low energy levels, sluggishness, and mood fluctuations. Its absorption and bioavailability after oral administration are under 1% increasing to 3% when taken in an *enteric (a coating that protects the capsule from the stomach acid)* capsule.

- SAMe antidepressant effects seem to be derived from its role as neuro-chemical in the *one-carbon cycle,* as it is able to donate methyl groups to neurotransmitters that affect mood, particularly *catecholamines. Catecholamines* are a family of chemicals to include *dopamine, adrenaline,* and *noradrenaline.* An enhanced methylation process can promote *dopaminergic* activity while also increasing the *serotonin* rate of replacement leading to an "uplifted" mood.
- In those with osteoarthritis, SAMe has been seen to provide relief as it further builds up in the body. Meaning its effects will not be felt until the SAMe levels have returned to within normal levels. SAMe seems to have this effect as it's involved in the synthesis of **glucosamine.** Glucosamine has been shown to have *chondroitin-protective (cartilage structural component)* and anti-inflammatory effects which can inhibit disease progression.
- SAMe anti-inflammatory effect seems to be derived from *Cyclooxygenase-2* inhibition; COX-2 has been implicated in inflammation and the aggravation of cartilage inflammation.

Proposed benefits (evidence varies)

- Anti-inflammatory
- Aid in osteoarthritis
- Aid in depression

Possible Side Effects

- Constipation
- Diarrhea
- Vomiting
- Sweating
- Dizziness
- Insomnia
- Gas
- Headache
- Anxiety

Dosages

- 260-530 mg 3x a day for osteoarthritis eventually going back down to 400 mg daily after 8 weeks

WARNING

Seek medical advice before use. SAMe can trigger mania in bipolar disorder and interact with serotonergic medicines, causing serotonin syndrome. Parkinson's disease treatment and Lesch-Nyhan syndrome require specialist advice because of potential neurotransmitter interactions. Do not replace prescribed depression care.

Salmon Oil

Salmon Oil, like all other fish oils, is high in **omega-3** fatty acids. Not much differs from fish oil to fish oil, except for a select few compounds, for example, *astaxanthin. Astaxanthin* can be found in **krill oil** and salmon oil; it is what gives salmon its pink pigment and is able to embed itself into cell membranes as it's fat-soluble, making it a top antioxidant. The omega 3-fatty acids found in Salmon Oil are *docosahexaenoic acid (DHA)* and *eicosapentaenoic acid (EPA),* which the body can make in limited amounts from ALA. *Omega-3* supplementation has been claimed to reduce the risk of *Cardiovascular Disease.* Ordinary supplements do not reliably prevent *cardiovascular events; selected prescription products have different evidence, and high doses can increase* atrial-fibrillation risk. *Astaxanthin* and *omega-3* show synergism benefits against inflammation and oxidation.

Docosahexaenoic (DHA) and *Eicosapentaenoic (EPA) Acids are* generally derived from various **fish oils, with no univers**al clinical superiority established for **krill or** Salmon Oil. *DHA* has nootropic effects that are further enhanced by *EPA's* ability to increase blood flow to the brain; this may very well aid those diagnosed with cognitive problems. It is important to note, those who regularly DO NOT eat fish as part of their diet, *DHA* may increase cognitive functions; for those that DO, benefits *<u>may not be seen from this supplement</u>*.

It has also been seen to reduce inflammation through the suppression of *phosphoinositide 3-kinase*, a signal transducer involved in inflammation. It can also reduce inflammation through modifying the production of eicosanoids; eicosanoids are messengers that are released after **arachidonic acids** are broken down; these messengers can either inhibit or cause inflammation.

- This may reduce plaque build-up in the arteries as it prevents *endothelium (the interior surface layer of vessels, i.e., blood/vascular cells)* inflammation, increasing cardiovascular health. This may be beneficial at reducing joint inflammation especially those with osteoarthritis.

Omega-3 fatty acids have been seen to have the ability to enhance protein synthesis as well as affect *myostatin* (*GDF-8*) abilities. *Myostatin*, in short, is a growth factor; simply put, myokine is a protein that's released by *myocytes* that directly affects the muscle cells' ability to signal the inhibition of *myogenesis* (creation of muscle tissue).

<u>Benefits</u>

- Promote joint health
- Anti-inflammatory
- Enhance cardiovascular health
- Enhanced cognitive function
- Antioxidant
- Promote muscle growth

<u>Possible Side Effects</u>

- Loose stools
- Fishy breath
- Upset stomach
- Nausea

<u>Dosage</u>

- 1 gram a day to include EPA and DHA generally 735 mg for EPA and 265 mg for DHA

WARNING

Seek medical advice before use. Salmon Oil may cause an allergic reaction to those with **Seafood/Shellfish allergies**. Salmon Oil can also increase the risk of bleeding in those with **clotting problems.**

Saw Palmetto *(Serenoa repens)*

Saw Palmetto, from the *Arecaceae* family, is a small palm tree that can be found in the Southernmost Eastern part of the United States. Its berries have been used in extracts or eaten in traditional Native American medicine for the treatment of various urinary problems in men. Currently it is sold and marketed as a testosterone booster, with mixed results, and as a treatment for *Benign Prostatic Hyperplasia (BPH).* The active compounds seem to come from the matrix of fatty acids and sterols that are found in the berries; high-quality trials have not shown urinary symptom benefit over placebo, and testosterone enhancement is unproven.

When used for BPH, it is believed that its main mechanism is the inhibition of *enzyme 5α-reductase. Enzyme 5α-reductase* is responsible for turning *testosterone* into *dihydrotestosterone (DHT).* In theory, the inhibition of DHT conversion would increase free flow testosterone in the body which would lead to enhancing *musculoskeletal (muscle)* development; in application, though, testosterone has been seen unchanged unless paired with other nutrients such as *astaxanthin* the antioxidant.

- DHT can stimulate prostate cell *proliferation (multiplication)*, leading to an enlarged prostate.

Oleic Acid, Myristic Acid and *Lauric Acid* found in Saw Palmetto have been seen to inhibit the *5α-reductase enzyme,* thereby reducing *DHT* in the body which can inhibit the growth of *BPH.* A reduction of *DHT* can inhibit *fibroblasts* and *myofibroblasts (involved in cell type change) proliferation (cell multiplication)* in the prostate.

- Fibroblast stimulation is the process in which fibrocytes make the matrix or "space" that's outside of the cell to enhance structural and biochemical support of the cells around it.

Urinary relief in men may be derived from an inhibition of the *α1-adrenergic receptor. The α1-adrenergic receptor* is responsible for *smooth muscle* contraction. Smooth muscle can be found in the stomach, brain, colon, and kidneys. *α1-adrenergic receptor* contracts the bladder as well as *ureters* that move urine from the kidneys to the bladder and the *urethral sphincter,* which control the exit of urine from the bladder into the *urethra.* If the *urethral sphincter contracts, the urethra* closes making urination difficult.

Proposed benefits (evidence varies)

- Aid with Benign Prostatic Hyperplasia
- Relieve urinary problems
- May increase testosterone w/astaxanthin

Possible Side Effects

- Gastrointestinal discomfort
- Dizziness
- Vomiting
- Constipation
- Headache
- Diarrhea

Dosages

- 160 mg 1x a day or 320 mg broken in 2 doses of a standardized 80-90% fatty/liposterol extract.

WARNING

Seek medical advice before use. Palmetto Oil may affect those with **hormone sensitive conditions** as Palmetto Oil can affect the conversion of hormones like **estrogen**, **estradiol**, and **testosterone.**

Selenium

Selenium is an essential mineral and an element that can be found in the periodic table of elements, and although it is toxic at high doses, it is necessary in trace amounts for proper cell function and as a *component of* antioxidant selenoproteins such as **glutathione peroxidases**. Generally, if you take a multivitamin, supplementation isn't necessary as your daily dose should be met. Any more than that can cause oxidative damage to the body.

Everyone that is currently reading this book needs Selenium in their body; however, Selenium deficiency is rare in industrialized countries. There are currently 25 identified *selenoproteins* found in the body, and as of current the role of the majority of these proteins is still unknown; the ones that are known have been seen to play a huge role in reducing oxidation in the body.

- *Selenoprotein S* protects the *endoplasmic reticulum* from stress. A function of the *endoplasmic reticulum* is to fold proteins and transport them into the *Golgi Apparatus*. If the *endoplasmic reticulum* is stressed, unfolded proteins are created, which result in *unfolded protein response; this* response is basically a cellular stress response. If this continues, it can lead and snowball into damage derived from hypoxia and can eventually lead to various disorders.

It also aids the antioxidant enzyme *thioredoxin reductase* in *REDOX* reactions.

- *REDOX* stands for *Reduction* and *Oxidation*. This is crucial for antioxidant and metabolic effects in the body. *Oxidation* is where an atom loses an electron and oxidizes, while *Reduction* is where an atom gains an electron and becomes less oxidative. The process can form a new compound; this is an important factor in the biological energy conversion of glucose to antioxidant activity in the body.
- Selenium interacts with *thioredoxin reductase* to affect DNA transcription through *REDOX*, as it has been seen to regulate gene expression that affects cell *proliferation (cell multiplication)* and *apoptosis (programmed cell death).*

Selenium's antioxidant role does not establish prevention of cardiovascular disease or cancer by supplements in adequately nourished people. Low selenium levels have been linked to Pre-eclampsia.

Benefits

Supports general health in the body

Possible Side Effects (Seen in high doses)

- Irritability
- Vomiting
- Garlic breath
- Brittle nails
- Sluggishness
- Loss of energy
- Hair loss
- Fatigue
- Nausea

Dosages

- Adult RDA: 55 mcg/day from all sources. US adult upper limit: 400 mcg/day, not a target or an extra dose; food intake varies.

WARNING*

Seek medical advice before use. Review **autoimmune disease,** including **rheumatoid arthritis** or **multiple sclerosis**, and **thyroid treatment because** selenium affects immune and thyroid functions. Excess causes selenosis; normal nutritional intakes are not established to worsen these diseases.

Sesamin

Sesamin is a *lignan (*a group of *polyphenols)* that is found in sesame seed oil or extracted from the *Fagara* plant found in the *Rutaceae* family where citrus plants can be found. It can also be referred to as *sesamum indicum* extract. It has been marketed and advertised as a fat loss/anti-obesity product, but being that it is a polyphenol, it can also be seen as an antioxidant as well. The seeds themselves are small flat and can vary in color from brownish to black.

Its main mechanism of function seems to act as an inhibitor to *delta-5-desaturase* enzyme and *tocopherol-omega-hydroxylation. Delta-5-desaturase* is involved in fatty acid metabolism, while *tocopherol-omega-hydroxylation* is involved in Vitamin E metabolism.

- By inhibiting *tocopherol-omega-hydroxylation*, Vitamin E oxidation is lowered and its levels can be increased throughout the body. Since Vitamin E, and its 8 variations, are seen as antioxidant and anti-inflammatory compounds, this can become beneficial to the body by reducing free radical damage while also reducing any inflammation that may have come from the free radicals. Additionally, Sesamin seems to have antioxidant effects on its own as it has the capability to remove *Superoxide* and *Hydroxyl* radicals from the body.
- Since Vitamin E is fat-soluble, it can be stored in fatty areas; fatty acid oxidation would be a secondary method of Vitamin E degradation. Since Sesamin is able to inhibit the *delta-5-hydroxylation* enzyme that is directly responsible for fatty acid metabolism/oxidation, it is believed this could additionally increase the circulating levels of Vitamin E in the body.

It is important to note as of current events, there has been no data or trials to confirm the marketed benefit of fat loss from this supplement. Although there are anecdotal reviews stating that it has worked, there currently has been no clinical data to confirm or deny this claim. The fat loss effect is believed to come from the downregulation of *Peroxime Proliferator-activator Receptor Alpha (PPARα)* since *PPARα* interacts with *adipose (fat)* tissue. It has been seen that *PPARα* can directly influence gene transcription in enzymes that are designed for and ultimately optimize the oxidation of fatty acids while redirecting fat storage.

- This can lead to a lower expression of *triglycerides*, promoting a favorable lipid metabolism.

Benefits

- Improve fat loss
- Lower cholesterol levels
- Anti-inflammatory
- Antioxidant
- Promote nutrient use

Possible Side Effects

For lack of clinical data and anecdotal reports, there currently are no known or reported side effects.

Dosage

- 750-1500 mg a day

WARNING

Please consult with a medical professional prior to consuming Sesamin. There is concern that Sesamin may affect those with **low blood pressure** as Sesamin can lower blood pressure. If you have **diabetes,** Sesamin may affect blood sugar levels as *PPAR Alpha* can affect insulin sensitivity, thereby affecting blood sugar levels and response.

Shark Cartilage

The basis of Shark cartilage as a supplement came from a book in 1992 called "Sharks Don't Get Cancer." Since that time the supplement has been advertised and marketed as a miracle health supplement, with claims of curing various diseases to even preventing cancers. As of current this supplement is still sold, but it is currently more scoped as a joint supplement rather than a heal all potion. With that being said, there currently is no clinical data to support any of its supported claims, and there have been no noted significant differences between Shark Cartilage and a placebo. Sharks do develop cancer. *Type II collagen* contributes to cartilage structure, but shark products are not proven identical or clinically equivalent to s*pecific undenatured-collagen preparations*.

Specific undenatured *Type II collagen* preparations have been studied for joint symptoms; these findings cannot simply be transferred to shark cartilage.

- *Type-II Collagen* is collagen that is found and used to create the cartilage found on the surface of joints. Immune responses to cartilage components can contribute to rheumatoid arthritis; excess collagen release is not a general explanation of arthritis.

An oral-toleranc*e mechanism* has been proposed for specific undenatured Type II collagen preparations as one would regularly see from the process mentioned above by creating a type of, "tolerance," to the level of *Type II Collagen* that is released in the body. It is able to accomplish this by affecting *dendritic cells*. *Dendritic cells* are the basis in response to *autoimmune inflammatory* diseases like rheumatoid arthritis. *Dendritic cells* are capable of producing immune responses, such as inflammation.

- When one consumes something that contains *Type II Collagen* with preserved relevant epitopes, the collagen interacts with *Peyer's patches* which are areas in the intestines with a high level of *dendritic cells*. With continued consumption, *dendritic cells* can change (become de-sensitized to collagen in the body in a sense) and induce a different response from T-Cells; a T-cell is a type of white blood cell (Elevated T-Cell concentrations are found in the synovial fluid in those with rheumatoid arthritis). This changed response is then able to indirectly influence the production of *interleukin 10 (IL-10)*. The release of *IL-10* inhibits the release of IL-17; IL-17 can drastically affect inflammatory diseases in a negative manner.

This response can create an anti-inflammatory environment while also reducing the pain that is brought on through inflammation. The overall mechanism is similar and has been compared to that of a vaccine.

Proposed benefits (evidence varies)

- Anti-inflammatory
- Promote joint health
- Anti-arthritis

Possible Side Effects

- Gastrointestinal discomfort
- Upset stomach
- Nausea
- High Blood sugar
- Dizziness
- Vomiting
- Low Blood Pressure
- Constipation
- Fatigue

Dosage

- A 40 mg/day studied undenatured Type II collagen dose is not an established equivalent dose for shark cartilage; cancer-treatment benefit is unproven.

WARNING

Seek medical advice before use. Shark Cartilage may affect those with **Hypercalcemia** as Shark Cartilage can increase calcium levels.

Spirulina *(Arthrospira platensis)*

Spirulina is a spiral-shaped blue-green algae derived from two types of algae, *Arthrospira platensis*, and *Arthrospira maxima*. The consumption by humans became popularized shortly after it was used by NASA as a supplement for astronauts. It has been marketed as an immune booster, anti-inflammatory, and a possible neurological antioxidant. Due to its high concentration of B-12, it has been used as a source for Vitamin B-12, but much is inactive B12 analogue; Spirulina is not a reliable source of active human B12. It is a good source of vegan protein since it is a complete protein.

The main active compounds seen in Spirulina are *phycocyanobilin proteins* which have been seen to have a similar molecular structure to *bilirubin* found in the human body.

- *Bilirubin* is involved in the destruction of aged blood cells (as they can become toxic) and has been seen to play a role as an antioxidant in the body.

A slight *bilirubin* elevation has been associated as beneficial as seen with Gilbert's Syndrome. Gilbert's Syndrome has shown slightly higher than average bilirubin levels; this appears to be protective as they have shown longer lifespans than those with average *bilirubin* levels. Since Spirulina has a similar molecular structure as *bilirubin,* it is believed that it can mimic Gilbert's Syndrome and promote antioxidant activities as it is able to inhibit *Nicotinamide adenine dinucleotide phosphate (NADPH) oxidase* in the same manner as *bilirubin.*

- *NADPH oxidase* is beneficial to the body in a regulated form as it creates *Superoxide* free radicals. Underactivity can cause susceptibility to microbes while overactivity can cause oxidative damage which can lead to various diseases like cardiovascular disease.

The same mechanism seems to be applicable to neurons and various other tissues found throughout the body. Interestingly enough, Spirulina has been seen to be able to be effective at reducing *Low-Density Lipoprotein (LDL)* Cholesterol as it has reduced absorption while increasing the amount that the body can defecate out. The mechanism towards this actually is still under scrutiny. GRAS conclusions concern specific preparations and uses, not all products. Microcystin or metal contamination can occur, and side effects are possible.

Proposed benefits (evidence varies)

- Enhance immune system
- Anti-inflammatory
- Neurons-protective
- May lower cholesterol
- Antioxidant

Possible Side Effects (These effects have been seen in toxic metal-contaminated Spirulina)

- Gastrointestinal discomfort
- Weakness
- Nausea
- Shock
- Vomiting
- Thirst
- Rapid heartbeat
- Erratic heartbeat

Dosage

- 200 mg per 2.2lb/kg per body weight

WARNING

Seek medical advice before use. Spirulina may affect people with **Phenylketonuria** as Spirulina contains **phenylalanine** which can make phenylketonuria worse. Those with **autoimmune diseases** should not consume Spirulina as it can stimulate immune responses to increase autoimmune disease symptoms and worsen conditions.

St. John's Wort *(Hypericum perforatum)*

St. John's Wort is a bright yellow flowering plant from the *Hypericaceae* family and has been used and marketed for its antidepressant effects. It is found in the majority of the world and is believed to have derived the name from John the Baptist as it blooms around June 24, St. John's Day, and when crushed a purple/reddish liquid comes out that is said to symbolize St. John's blood.

The main active components that have been found in St. John's Wort are Hypericin and *Hyperforin. Hyperforin is more bioactive and has* been seen to be a psychoactive compound and seems to interact with *catecholamines*. The suggested action is believed to induce an influx of ions into *non-selective cation channels (NSCC)* that are found in neurons. An increase of ions, particularly sodium, can inhibit the uptake of *serotonin* thereby increasing the free flow of *serotonin* in the body; this is done selectively and does not affect the *Serotonin Transporter* as seen in antidepressant medications.

- *Serotonin* is a neurotransmitter that has been associated with feelings of happiness and well-being.

This mechanism is believed to occur to other transmitters as well, such as *dopamine* and **GABA**.

- *Dopamine* plays a role in reward-motivated behavior. While GABA is responsible for sedation and relaxation, this is responsible for reducing anxiety and sleep as well.

An *extracellular (outside of the cell)* increase of these neurotransmitters in depressed individuals can be associated with a reduction of depression symptoms.

***St. John's Wort is pharmacologically active. Certain standardized extracts have evidence in mild-to-moderate depression, but benefit in severe disease is uncertain. It induces drug-metabolizing enzymes, including CYP3A4, and P-glycoprotein, reducing exposure to many medicines. Medical interaction screening is essential; do not adjust prescribed treatment yourself.

Proposed benefits (evidence varies)

- Promote a feeling of well-being
- Antidepressant
- Possible anxiety relief

Possible Side Effects

- Insomnia
- Vivid dreams
- Anxiety
- Irritability
- Upset stomach
- Dizziness
- Skin tingling
- Rash
- Diarrhea

Dosage

- Trials commonly use standardized extracts around 300 mg three times daily; products differ. Interaction screening and professional guidance are essential, not a 1–2 *mg general regimen.*

WARNING

Seek medical advice before use. St. John's Wort can trigger mania in bipolar disorder and worsen psychosis; use with neurological conditions, including Alzheimer's disease or **ADHD**, requires individual review. It can reduce effectiveness of contraceptives and transplant, HIV and cancer medicines. Serotonergic combinations can cause serotonin syndrome. Tell the surgical team well in advance: enzyme effects can persist after stopping, and the team should set the stopping interval.

Stevia *(Stevia rebaudiana)*

Stevia is generally referred to as the sugar substitute that is extracted from the Stevia plant. The plant is part of the *Asteraceae* family where the sunflower is found; it has extremely sweet leaves, and the extract has been found to be anywhere from 250-300 times sweeter than sugar. The plant has been traditionally used by the Guaraní People of South America to sweeten tea and at times some medicines. Currently it has been used and marketed as a healthier sugar substitute and has been used by some companies to replace sugar in their products. It is also a zero-calorie sweetener as *steviol glycosides* cannot be metabolized by the body.

Steviol glycosides are able to maintain their structure throughout the intestines, until finally in the colon *steviol glycosides* are then broken down. Bacteria there are able to *hydrolyze (separation of molecules by consumption of a water molecule) steviol glycosides* and break off the glucose molecule. *Steviol* is then able to be absorbed and metabolized in the liver to form *steviol glucuronide*. This compound is then able to be removed by the body through urine.

- Interestingly enough, none of the byproducts or molecules of stevia are stored in the body; they generally pass through the body without interacting with it. The glucose molecule is so minuscule that absorption via the body causes an effect that it can't even be classified as minimal.

Various proposed beneficial and health effects are very insignificant and are generally acquired at high doses which can lead to toxicity in other parts of the body. The main health benefit comes from the replacement and/or reduction of glucose generally derived from one's diet, such as soft drinks and sugary foods. A natural origin does not establish superiority to other authorized sweeteners; purified steviol glycosides and crude whole-leaf products have different safety and regulatory treatment.

Currently, however, bioactive compounds from the plant, the non-sweetener parts, have been looked upon as powerful antidiabetics. They have been seen to augment *glucose-stimulated insulin secretion*, meaning the extract is only able to stimulate an insulin response in the presence of glucose. Mainly by augmenting the *muscarinic insulinotropic (insulin release stimulated via the acetylcholine neurotransmitter)* effects, mainly through the *Muscarinic Acetylcholine Receptor 3 (regulates insulin secretion)* in the *pancreatic islet (areas in the pancreas that contain endocrine/hormone cells)*.

Proposed benefits (evidence varies)

- Antidiabetic
- Zero Calorie sweetener

Possible side Effects

- Bloating
- Numbness
- Nausea
- Muscle Pain

Dosages

- ADI: 4 mg/kg/day as steviol equivalents, not extract weight. Purified glycosides differ from crude leaf products; this safety ceiling is not a target.

WARNING

Seek medical advice before use. Stevia may cause an allergic reaction if you are allergic to **Ragweeds** as Stevia and ragweeds are part of the same family, Asteraceae.

Strontium

Strontium is a chemical element found on the Periodic Table sitting at atomic number 38. It is in the Alkaline Earth Metal group where calcium and magnesium are found, and it is a malleable silver-yellowish metal. Before going further into this "supplement," it is important to make a note: Strontium *intake has been associated with severe side effects; supplementation is not recommended without the supervision of a licensed doctor or medical professional. Long-term prescription studies do not establish safety of all strontium salts*. As such the only studies that have been done on Strontium involve the medical prescription drug *Strontium ranelate* under the name Protelos ®. Much of Strontium's mechanism will be of Strontium ranelate as of current there is no clinical data for basic Strontium supplementation.

When looking at strontium to calcium ratio in the bones, it is found to naturally occur at a ratio of 1:1000-2000; any higher than that, bone development can be affected in younger people such as kids. Since Strontium is very similar to calcium, it can be absorbed in a similar manner and interact with the body in a similar fashion as calcium. Strontium's main purpose is that it would displace calcium in the bone, being that it is a heavier molecule weighing in at 87.63 u compared to Calcium which weighs at 40.078 u, it would, in theory, increase bone density.

- u is the Unified atomic Mass Unit is the molecular scale that quantifies atomic mass.

Strontium absorbs X-rays more strongly than calcium, artificially raising DXA bone-density readings. This measurement effect does not by itself demonstrate stronger bone or fewer fractures.

Strontium ranelate is able to break the bone remodeling process, by enhancing *osteoblast (bone creation)* and inhibiting *osteoclasts (bone break down)*. By increasing/enhancing bone cell creation and reducing bone cell *apoptosis (programmed cell death)*, bones can be re-mineralized, and the addition of strontium can increase bone density leading to fewer fractures and breaks in those with weaker bones such as menopausal women with *osteoporosis*.

Proposed benefits (evidence varies)

- Increased bone density
- Improve osteoporosis symptoms

Possible Side Effects

- Dermatitis
- Diarrhea
- Nausea
- Headache

Dosage

- The 2 g/day regimen studied prescription strontium ranelate, not routine over-the-counter strontium salts. Osteoporosis treatment requires specialist assessment.

WARNING

Seek medical advice before use. Prescription strontium ranelate has cardiovascular and thromboembolic risks; findings cannot be transferred to over-the-counter salts. Avoid self-treatment with **clotting** disorders, prior **stroke**, **peripheral arterial** disease, **myocardial infarction** or other cardiovascular disease. Venous thromboembolism and pulmonary embolism are concerns. A DXA increase can reflect X-ray absorption rather than stronger bone; discuss osteoporosis treatment with a clinician.

Sustamine ® *(L-Alanyl-L-Glutamine)*

Sustamine is a dipeptide molecule made up of the two amino acids, **L-Alanine** and **L-Glutamine**. Both of these amino acids have properties that are synergistic with each other and can be used in both the medical field for things such as post-operative surgery, gastrointestinal disorders, trauma, starvation and burn victims, as well as athletic performances for hydration and energy. The reason why it is so effective in the medical field is because it is a dipeptide and it enhances its stability and absorption.

Generally speaking, L-Alanine-L-Glutamine works in the same manner as L-Alanine and L-Glutamine; the benefit of them paired together is the increased absorption rate and stability. As two separate amino acids, absorption rates would be slower as the amino acids would have to wait for transporters to get through the intestinal wall to be used by the body.

- Most transporters in the body can only carry one amino acid or a dipeptide at once. *Pept-1 (peptide transporter in the intestines)* transporter is able to transport the dipeptide for a faster absorption rate compared to transporting one amino acid at a time.
- L-glutamine also degrades quickly after being introduced to liquids and is also temperature sensitive. Glutamine levels have been seen to be increased in the body, and in muscle tissue; the prominence of enhanced athletic performance has not yet been observed. Also, glycogen levels have not been seen to increase in muscle tissue any more than regular glutamine supplementation.

For medical purposes, this is beneficial for water and electrolyte absorption in cases of diarrhea or vomiting. It is also able to increase *heat shock proteins 72* and *73* expression. These proteins protect cells from metabolic and oxidative stress; they are able to refold and stabilize mutated proteins and aid with DNA repair.

Any performance-based benefits seen with L-Alanine and L-Glutamine supplementation are the same as Sustamine with no significant difference from one or the other. Please refer to the supplement **L-Alanine** and **L-Glutamine** for a full scope of performance benefits that are applicable to Sustamine. Any "enhanced" benefit that has been observed in Sustamine is for medical applications of the dipeptide, where singular amino acids do not show the same efficacy.

Benefits

- Improved hydration and electrolyte absorption in post-operative surgery, gastrointestinal disorders, after trauma, starvation and burn victims.

Possible Side Effects

Side effects may be similar to or a combination of side effects seen in both L-Glutamine and L-Alanine. Please refer to those pages for more detail.

Dosage

- 1-3 g a day, usually prior to a performance event to increase hydration and electrolyte absorption

WARNING

Please consult with a medical professional prior to consuming Sustamine. There is concern that Sustamine may complicate conditions of any person that is affected by either L-Glutamine or L-Alanine; please refer to those pages for an in-depth warning and possible side effect explanation.

Synephrine *(Bitter Orange/Citrus aurantium extract)*

Synephrine, also called and listed as bitter orange extract, is an *alkaloid* molecule that has similar but weaker mechanisms as *ephedrine (stimulant medication)*. Synephrine is derived from the unripe, dried oranges from the *citrus aurantium* family and has long been used under the name Zhi Shi in traditional Chinese Medicine. There are various names that are used for synephrine such as *m-synephrine* and *o-synephrine*, and though they all have similar structures and mechanisms, *p-synephrine* is the main extract found in *citrus aurantium.* It is mainly seen marketed as a weight loss supplement seen in products such as Zantrex and as a performance-enhancing product commonly seen in some pre-workouts. Weight-loss efficacy and safety remain unclear; cardiovascular events are reported, often with multi-ingredient products.

Synephrine's main mechanism of function comes from being a *beta-agonist.*

- A *beta-agonist* is a compound that is able to relax and widen muscles, especially those involved in breathing such as airways. They can also mimic the same effects that are seen with *epinephrine* and *noradrenaline* in smooth and cardiac muscle such as the heart, stomach and lung function.

Beta-agonist activation can lead to increased *basal metabolic rate (metabolism rate at rest)* and *lipolysis (metabolism of fat cells for energy).*

- Activation of *Beta-agonist β3* can induce *adipose tissue (body fat) lipid* metabolization. This effect can promote fat loss over a period of time.
- Activation of *Beta-agonist β2* can induce relaxation of blood vessels while increasing heart muscle contractions and rate. (This can be beneficial at the time of peak physical performance but can also cause heart problems in individuals with a history of heart problems.)

Combining synephrine with caffeine may amplify stimulant and cardiovascular effects; this is not proof of safe or effective weight loss. Since caffeine can also interact with *catecholamines (like dopamine and epinephrine),* it has been shown to have more pronounced effects when taken together.

- Caffeine is also able to reduce the amount of fatty acids absorbed by the intestines while also temporarily inducing insulin resistance and reducing glucose disposal without affecting current insulin levels. This can create an enhanced fat loss, especially if complemented with exercise.

Proposed benefits (evidence varies)

- Improve athleticism
- Enhance fat loss
- Lifted mood

Possible Side Effects

- Increased heart rate
- Headaches
- Migraines
- High blood pressure
- Abnormal heart rhythm
- Irregular heartbeat

Dosage

- 10-20 mg 3x a day or one 50 mg dose a day

WARNING

Seek medical advice before use. Synephrine may complicate conditions of any persons who have a history of **heart** and **blood pressure** problems such as **heart arrhythmia, heart disease** and **high blood pressure**. Synephrine interacts with neuro-modulators that affect heart rhythm and blood pressure that may worsen the present condition.

Theobromine *(3,7-dimethylxanthine)*

Theobromine is a *xanthine alkaloid* like caffeine and has a similar chemical structure like caffeine; it differs mainly from the fact that Theobromine has two methyl groups, while caffeine has three; this importance will be explained in the second portion of this page. *Theobromine is one of the main alkaloids found in chocolate, and for that reason at times theobromine and its bioactive compounds will be referred to as cocoa extract on some labels.* It is important to note that although the name is Theobromine, there is no *bromine (element 35 in the periodic table)* within the compound; the name is derived from Theobroma genus where the cocoa tree is found.

Although Theobromine is structurally similar to caffeine, it is generally less centrally stimulating than caffeine; no universal percentage potency applies. Differences in receptor affinity and metabolism matter; methyl-group count does not establish a simple potency ratio. When compared to caffeine, its half-life is 2.5-5 hours and reaches peak levels in the body at about 30-40 minutes. Theobromine, on the other hand, has a half-life of 7-12 hours but also takes 2-3 hours to reach peak levels in the body. Caffeine can more readily cross the blood-brain barrier in comparison to Theobromine.

Theobromine does not seem to provide any effect on the body by interacting through the central nervous system like caffeine; instead, the majority of its effects seem to come from relaxation and *vasodilation* in the body. It also does not provide an alert feeling like caffeine does; instead, it also promotes a feeling of relaxation and the good feeling most people associate to eating chocolate. The stated effects tend to differ from individual to individual as genetics play a role in *xanthine* sensitivity. Its main mechanism seems to be derived from *phosphodiesterase* inhibition.

- By inhibiting *phosphodiesterase* enzymes, it seems to have an antidepressant effect as this prevents *phosphodiesterase* enzymes from converting *cAMP* to an inactive form. *cAMP* has a role as a messenger of hormone and neurotransmitter-controlled systems, involving mood regulation all the way up to inflammation and vasodilation.

Proposed benefits (evidence varies)

- Improved mood
- Vasodilator
- Bronchodilator

Possible Side Effects (These effects are typical due to high doses of 1200 mg and up)

- Nausea
- Anxiety
- Insomnia
- Headaches
- Negative mood changes
- Loss of appetite

Dosage

- 25-40 g of 85% dark chocolate to receive beneficial amounts of Theobromine.
- 500-1000 mg a day for mood enhancements

WARNING

Seek medical advice before use. Theobromine may affect people with **bleeding disorders** as Theobromine may increase bleeding and clotting time. If you have **low blood pressure** problems, theobromine is a vasodilator and can lower blood pressure to unsafe levels. In those with reflux disease, Theobromine can worsen conditions as it can relax smooth muscle in the esophagus leading to acid reflux. Theobromine can also affect those with **heart conditions** such as **tachycardia** as it is seen to interact with its signaling.

***Tribulus terrestris* (Devil's Weed)**

Tribulus terrestris is part of the *zygophyllaceae* family and can be found growing throughout the world; in the United States it is considered a weed and an invasive species. *Tribulus terrestris* can go by a few names such as Devil's Weed, Goat's Head (since it looks like a goat) and various others. It can survive in just about any weather and has been seen to have been used in traditional Indian Ayurvedic medicine and some documentation shows traditional Chinese Medicine use as well. It was generally used to treat urinary problems and infections as well as improve male sexual function and libido. Currently it is marketed as a testosterone booster and can be seen as a part of testosterone boosting product formulas, although controlled human evidence does not establish increased testosterone or muscle-building benefit.

The main bioactive compounds in this plant seem to be *steroidal saponins*, particularly *protodioscin* and *protogracillin*, which are where the claimed muscle-building benefits and the increased libido come from. As with many supplements the species and area that it's cultivated differ in potency; Turkey's strand, *Ankara* being the most potent. It is believed that *protodioscin* and *protogracillin* are able to achieve this effect through an enhanced/increased conversion of testosterone into the more potent *dihydrotestosterone (DHT).*

- *DHT* has been associated with a high anabolic activity coming from increased blood circulation and production along with an increased sex drive.
    - *DHT* is involved in the development of male differentiation in embryos, as well as the defined male attributes seen in males at puberty.

The increase in blood flow seems to be independent of *Nitric Oxide* activity in which blood *vasodilation (widening of blood vessels)* is generally derived from. The increased blood flow will allow for increased circulation in contracted tissue, aiding in proper erectile function. The current mechanism of erectile aid is unknown, and it is believed to be tied to *Androgen Receptors (AR)* stimulation. The *AR* stimulation is based on an immune reaction. Particularly put that *Tribulus terrestris* molecules can stimulate the immune system in a manner that promotes the secretion of additional male hormones.

- *AR* has been shown to interact with various immune reactions/responses derived from immunoreactive proteins and *ligands.*

Proposed benefits (evidence varies)

- Possible testosterone booster
- Improve Erectile function
- Antioxidant

Possible Side Effects

- Stomach Pain
- Cramping
- Diarrhea
- Menstrual Bleeding
- Constipation
- Excitation
- Aggression
- Insomnia

Dosages

- No established testosterone- or muscle-building dose exists; experimental weight-based amounts are not suitable human recommendations.

******WARNING******

Seek medical advice before use. Tribulus terrestris can affect **diabetics** as it can lower blood glucose and increase insulin. Additionally, those at risk for **Prostate Cancers/Growth, the increased** levels of DHT can worsen conditions.

Turkesterone

Turkesterone and Ecdysterone are *ecdysteroids;* relative laboratory potency does not establish **human effectiveness**. *Ecdysteroids* are insect *steroid hormones* that are involved in molting and in the reproduction of insects. They are also found in some plants that create *ecdysteroids* in response to environmental stress. Currently they are marketed to enhance muscle mass while also increasing general health in humans. *Ecdysteroids* are structurally similar to *androgens* but this does not establish steroid-equivalent benefits, absence of hormonal effects or *long*-term human safety.

Preclinical studies describe enhanced muscle protein synthesis, but do not establish steroid-equivalent effects in humans. A 2025 randomized trial found no body-composition benefit with the studied supplement.

- There are various mechanisms in which it is able to accomplish this. One being through enhanced *mRNA Translation* directly geared towards enhanced protein synthesis, since protein synthesis is regulated during transcription and translation of RNA. This has been seen to happen in the *ribosomes (location in the cell where protein synthesis happens)* of *eukaryotic (cells with a nucleus and membranes)* cells.

Protein synthesis comes from *RNA Translation,* which uses an *mRNA* sequence as a template (kind of a like a recipe/instructions) to create a specific amino acid chain (the product it wants to build) based on its genetic code (rules for protein synthesis to create the specific product), in this case, it produces/enhances specific peptides/polypeptide used in protein synthesis needed for muscle repair.

- Second pathway has been seen to cause a Calcium ion influx into *myocytes (muscle cells)* which can cause *phosphorylation* of *AKT,* which ultimately signals protein synthesis. The *AKT pathway* promotes cell growth and responds to *extracellular (outside of the cell)* signals, such as a Calcium ion influx.
- The third pathway is based on the belief that ecdysteroids can increase *myotubes (muscle fibers)* size through interacting with the *α* and *β Estrogen Receptor* as activation of these receptors has been linked to an increase of *MyoD (a protein involved in muscle tissue formation).*

Additionally, *Ecdysteroids* are able to support *neurons* by protecting them against *neurological* toxins as it can promote the secretion of *acetylcholinesterase* and *glutamic decarboxylase.* These are enzymes in charge of breaking down over produced neurotransmitters that can cause toxicity to neurons.

Proposed benefits (evidence varies)

- Can reduce blood sugar
- Promote lean muscle
- Aid fat loss
- Does not affect hormones
- Enhance endurance
- Antioxidant

Possible Side Effects

There currently have been no reported side effects

Dosage

- No effective, safe human bodybuilding dose is established.

WARNING

Seek medical advice before use. Although there have been no known negative side effects reported, research is still lacking. If you have a **metabolic disorder,** seek medical advice before taking any ecdysteroid.

Turmeric (Indian saffron)

Turmeric is a plant that belongs to the *Zingiberaceae* family, where ginger is found and is native to Southeast Asia. Turmeric has been used as a spice in various Asian cuisines and has an earthy black peppering taste and looks like an orange-ish yellow powder when ground up. It also has various ties to traditional Indian and Chinese medicine, particularly Ayurvedic medicine. Recently it has been marketed as a powerful antioxidant and anti-inflammatory.

The main bioactive compound found in Turmeric is curcumin; due to its poor bioavailability *piperine (black pepper extract)* has been studied to inhibit *metabolism* and increase exposure; a small study reported a 2,000% bioavailability increase, not inhibited absorption. *Glucuronidation* is the process in which substances like drugs, hormones, etc. are metabolized or broken down. Curcumin is a *diarylheptanoid* which is a natural phenol; these are made by the plant to protect it from environmental stress. To humans they are antioxidants in the body.

- Curcumin is able to hinder *Activator Protein 1 (AP-1)* released by tumor promoters; making compliant *AP-1 (involved in cell growth and death)* production while enhancing *phase II enzymes (antioxidant enzymes that break down carcinogens).*
- Curcumin is able to suppress tumor cell signaling by enhancing the confirmation of *AP-1* signals. In short, it suppresses signals that are "off or weird" in a sense. It can also enhance the liver's ability to metabolize foreign substances such as antibiotics and alcohol.
- Curcumin is able to protect against glutamate *excitotoxicity (nerve cell death by over-stimulation)* by promoting or preserving *neurotrophin (proteins involved in neuronal survival, function, and development)* levels. The same effect may prevent *corticosterone (stress hormone)* induced neuronal cell death.

Curcumin has been seen to have a significant increase in *nitric oxide* levels found in the blood; this can lead to a similar vasodilation effect that's similar to those who exercise every other day. Additionally, it is able to quicken muscle recovery time by inhibiting *Atrogin1/MAFbx expression; abnormal Atrogin1/MAFbx* expression in *myotubes (muscle fibers)* leads to muscle tissue degradation.

Proposed benefits (evidence varies)

- Anti-inflammatory
- Antioxidant
- Neuro-protectant
- Vasodilator
- Improve muscle growth
- Anti-cancerous

Possible Side Effects

- Gastrointestinal discomfort
- Diarrhea
- Dizziness
- Low blood pressure
- Nausea

Dosages

- 3-4 grams a day, preferably with *piperine (black pepper extract)* or a dash of black pepper.

WARNING

Seek medical advice before use. Turmeric may affect people with hormone sensitive conditions such as **breast cancer and uterine** cancer as high doses of Turmeric can have an estrogenic effect in the body; additionally, it can cause DNA damage and **immune suppression**. In those with diabetes, Turmeric can lower blood sugar levels; medications may need to be modified. If you are diagnosed with **Gallbladder** problems, curcumin might increase bile production and make symptoms worse. Liver injury is reported with some highly bioavailable products; piperine can change drug exposure and is not a required safe addition.

Tyramine

Tyramine is a naturally occurring *trace amine (a family of neuromodulators)* that is derived from **tyrosine**. Tyramine is taken up by *norepinephrine reuptake transporter (NET)* which then induces *catecholamines* expression, but since it is not able to cross the blood-brain barrier, its effects are only limited to *peripheral* nervous system *sympathomimetic* effects. These effects will only affect the body, not the brain or the *central nervous system; effects* that can be observed would be things like headaches, migraines, hypertension, etc.

- *Catecholamines* are a family of chemicals to include *dopamine, adrenaline,* and *noradrenaline.*

Tyramine's effects are amplified with *MAOI (Monoamine oxidase inhibitor)* medications. These are the reasons as to why you may get headaches or migraines or potentially fatal hypertensive crisis after eating cheese or drinking wine if you are taking *MAOI.*

- *MAO,* in nerve terminals, are enzymes that are involved in the breakdown or metabolization of *amines* like tyramine. They can also be found attached to the mitochondria of most cells in the body. Generally, the amount of *MAO* found throughout the body will metabolize tyramine before it can cause any prominent side effects like a headache, but the addition of an *MAOI* will allow tyramine to be unchanged allowing higher levels to be absorbed into the body thereby amplifying the effects.

For the average individual, and for the majority of results seen through research, Tyramine supplementation is not needed or recommended as the majority of people, if not all, already consume enough tyramine through their diet; even if one consumes protein shakes, the amino acid can be found on the side of the container. With that being said, there are a few cases that state that tyramine deficiency or deficient *endogenous (created internally)* production of tyramine is linked to some cases of depression. It is important to note that this recent find is fairly new; the preliminary findings are barely in their infancy. It is important to discuss this with your doctor prior to supplementing with tyramine as high dosed tyramine consumption can cause severe and possibly fatal side effects.

Proposed benefits (evidence varies)

- May be beneficial for those suffering from depression

Possible Side Effects (Generally seen with high doses or with an MAOI)

- Severe headaches
- Nausea
- Vomiting
- Hypertension
- Chest pain
- Rapid heartbeat
- Anxiety
- Shortness of breath
- Nose bleed

Dosages

- With an MAO inhibitor, follow its specific low-tyramine diet and medicine guide; no single intake threshold guarantees protection from crisis.
- No established supplemental indication or universal migraine-prevention dose exists; tyramine is a trace amine, not an essential dietary amino acid.

WARNING

Seek medical advice before use. Tyramine may affect people with **blood pressure** or **cardiovascular** problems such as **hypertension** or a history of strokes as tyramine can increase the severity and frequency of hypertensive episodes which can lead to **brain hemorrhage** or increase the likelihood of a stroke.

Ubiquinol/Ubiquinone

Coenzyme Q10 cycles between oxidized ubiquinone and reduced ubiquinol; a se**mi**quinone intermediate also participates in electr**on transfer. CoQ10** is the name for the compound, not a separate third supplement form. Cells interconvert these states as electrons move through mitochondrial reactions. Oral absorption depends on formulation, food and the individual. Ubiquinol may produce higher blood levels in some comparisons, but a universal advantage after age 40 and a fixed 95% whole-body proportion are not established. A biochemical role in energy production does not guarantee more energy or better athletic performance from extra supplementation.

CoQ10, ubiquinol, and ubiquinone are fat-soluble vitamin-like substances found in any "breathing" cell that helps the body produce energy through ATP. The majority of the body's energy is created this way which is why you will find the highest amounts of CoQ10 in the heart, brain, liver, and kidneys. *Statins inhibit the Mevalonate Pathway shared by cholesterol and CoQ10 synthesis; this does not establish routine symptom relief from supplementation, and not all blood-pressure medicines block this pathway.*

- CoQ10 and its derivatives interact with the *electron transport chain* within the mitochondria to promote **ATP** production, where its energy-enhancing benefits derive.

Supplementation of CoQ10 or its derivatives can provide protective benefits and proper function in the organs listed above. CoQ10 seems to protect the *endothelium (lining of the circulatory system)* from *LDL cholesterol* by inhibiting *NF-kB* derived inflammatory effects. Additionally, it is able to combat oxidative damage through *REDOX (loss or gain of an electron in a molecule)* in practically every cell in the body while also maintaining Nitric Oxide activity through the same mechanism, leading to vasodilation aiding cardiovascular health along with providing athletic benefit for athletes.

- Low CoQ10 has been studied in fibromyalgia and heart disease, but association does not establish a cause or cure. Supplementation of CoQ10 has been shown to increase concentrations in heart tissue, which can help reduce damage to the heart.

Proposed benefits (evidence varies)

- Cardiovascular health aid
- Possible antidepressant
- Promote athleticism
- Promote energy
- Vasodilator
- Antioxidant

Possible Side Effects

- Nausea
- Loss of appetite
- Stomach discomfort
- Rash
- Diarrhea
- Dizziness

Dosages

- 100-200 mg of Ubiquinone a day
- Ubiquinol and ubiquinone doses are formulation- and indication-specific; age over 40 does not establish universal superiority.

WARNING

Seek medical advice before use. Any form of CoQ10 may cause allergic reactions in some individuals. If you are diagnosed with Diabetes, CoQ10 and its derivatives can increase insulin levels; medication may need to be modified. Review warfarin and insulin interactions; evidence does not establish every listed therapeutic or athletic benefit.

<u>Ursolic Acid *(3-beta-3-hydroxy-urs-12-ene-28-oic-acid)*</u>

Ursolic acid is a phytochemical found in the *epicuticular wax (wax that covers the outer surface)* of some plants such as apples. It can be found in strong concentrations in plant peels. It has a history of being used in traditional Ayurvedic Medicine, under the name Holy Basil, for things like diabetes and to increase libido. Currently it has been marketed for muscle growth and fat loss. <u>It is important to note that there are currently limited human trials and bioavailability/absorption is extremely low in humans.</u>

Ursolic Acid has been seen to reduce fat and increase muscle during consistent food intake since it is able to promote fat use for energy while also maintaining muscle integrity in periods of fasting. It is believed to do this by countering muscle breakdown through promoting *IGF-1* expression, additionally activating *PGC-α (master regulator of mitochondria, this can increase energy)* in muscle tissue. These effects, however, do not affect muscle cell *proliferation (multiplication); rather,* it increases muscle cell size. It has also been seen to promote protein synthesis with the addition of a protein-rich meal; it is suggested that this mechanism comes from enhanced absorption of amino acids.

- Interestingly enough, it seems to be able to induce protein synthesis through an additional synergistic manner, insulin-induced *phosphorylation* of AKT. The *AKT pathway* promotes cell growth and survival; it responds to *extracellular (outside of the cell)* signals, such as insulin.
    - Insulin also acts as a messenger to transport nutrients into muscle tissue.
    - Dependent on the signal, phosphorylation can activate or deactivate protein enzymes which means *phosphorylation* of *AKT* can ultimately signal protein synthesis.

Ursolic Acid is able to increase *brown fat* and decrease *white fat* in the body. Regular fat (*white fat*) stores calories while *brown fat* is mitochondria packed and is used to burn energy and produce heat. An increase in *brown fat* and muscle tissue can play an anti-obese role as these increases can lead to higher energy expenditure.

- Although the reason as to how Ursolic Acid increases *brown fat* is not fully known, it is believed that one of the mechanisms of action is through directly stimulating *brown fat* synthesis. Since skeletal muscle and *brown fat* have the same precursors and molecular mechanisms, insulin and IGF-1 can signal for *brown fat* creation.

<u>Benefits</u>

- Promote muscle growth
- Promote fat loss

<u>Possible Side Effects</u>

As of current, there are no known or self-reported side effects

<u>Dosages</u>

- 2-6.5 mg per 2.2 lbs/1 kg of body weight, the amount is broken in 3, spread throughout the day

<u>***WARNING***</u>

Please consult with a medical professional prior to administering Ursolic acid. There is concern that Ursolic Acid may affect those diagnosed with **diabetes** as Ursolic acid interacts with insulin which may affect glucose levels; medication may need to be modified. <u>It is important to note that high doses of Ursolic acid can damage DNA and can directly induce cell death. It has also been seen to reduce fertility in males.</u>

Valerian *(Valeriana officinalis)*

Valerian, also known as valerian root, is a flowering plant from the *Caprifoliaceae* family where many ornamental garden flowering plants can be found. The plant has white-ish to pink flowers that have a light sweetened scent to them and are native to parts of Asia and Europe. Valerian root has been used in traditional medicine in Ancient Greece. Currently, the extract and herbal tea of this plant are marketed as a sedative and an anxiolytic that has been associated as a sleep aid as well as an aid for *dysmenorrhea (menstrual cramps)* symptoms in women. It is important to note that many studies have shown no significant benefit in sleep quality, although anecdotal reports have claimed a deeper state of sleep.

The main bioactive compound found in Valerian is *Valerenic acid* which is found in high concentrations in the root. *Valerenic acid* seems to interact with **GABA-A** receptors as GABA plays a role in *anxiolysis (prevents anxiety)* and sedation.

- *Valerenic acid* seems to positively influence GABA receptors by enhancing GABA neurotransmission/signaling strength. This may be the reason as to why higher doses seem to have sedation/anxiolytic effects. Interestingly enough though, one of the derivatives found in Valerian, *acetoxyvalerenic acid,* can block the GABA-A receptor channel thereby inhibiting *Valerenic acid* benefit.
    - If possible, an extract of a ratio of 1:1 of *Valerenic Acid* to *acetoxyvalerenic acid* is recommended to promote valerian sedative/anxiety effect. The higher the amount of *Valerenic acid* to *acetoxyvalerenic acid* the more *anxiolytic* effect that can be felt.

Valerian has been seen to reduce menstrual cramps *(Dysmenorrhea)* as it can positively modulate potassium channels while also inhibiting calcium channels; a high calcium influx to muscle tissues has been linked with muscle spasms which may be the reason for its anti-spasmolytic benefits.

Licorice is synergistic with valerian and has been seen to enhance valerian's *anxiolytic* properties. Valerian may also help with *OCD* as it can reduce symptoms as some have been linked to anxiety.

Benefits

- Reduce anxiety
- Help fall asleep
- Sedative
- Reduce menstrual cramps
- Anti-spasm
- Aid with OCD

Possible Side Effects

- Sluggish
- Upset stomach
- Uneasiness
- Vivid dreams
- Headache
- Drowsiness

Dosage

- 250-450 mg an hour before bed; higher than normal doses can cancel out the sedation effect
- 250-300 mg 3x a day for anxiety or OCD

WARNING

Please consult with a medical professional prior to administering Valerian. There is concern that Valerian may affect those who are currently taking **sleeping aids** like benzodiazepines or diphenhydramine. Valerian can also enhance the depressant effects of **alcohol**.

Vanadyl Sulfate (Vanadium)

Vanadyl sulfate is a stabilized collection of vanadium compounds. Under the periodic table, Vanadium is a transition metal under the atomic number 23. Vanadyl sulfate was popularly used throughout the '90s in bodybuilding as a way to increase fullness and size in muscles since it has insulin-like effects. Although it can still be found, it is seldom used in modern bodybuilding as further studies showed possible toxicities. Currently Vanadyl sulfate is being looked upon as a possible aid in diabetics.

The mechanism of Vanadyl Sulfate is by prolonging *Insulin receptor substrate 1 (IRS-1)* signaling as *IRS-1* signals for insulin. A Degradation of *IRS-1* can reduce or limit insulin signaling which would limit the amount of glucose that can be taken up by the body. It accomplishes this by inhibiting *tyrosine phosphatase (PTP)* enzymes. *PTP* enzymes are generally stimulated by insulin release; the more insulin that gets released, the more *PTP becomes activated. PTP* then negatively regulates *IRS-1* by suppressing it and degrading it. By maintaining *IRS-1,* insulin signals can be prolonged.

This *mechanism* was used by bodybuilders as it can increase nutrient delivery into muscle tissue, but currently, this mechanism seems to be more beneficial to those with diabetes as it may increase insulin sensitivity.

- In diabetics, increasing and maintaining insulin levels can increase glucose uptake; this can lead to lower blood sugar levels. This mechanism may continue past the use of Vanadyl Sulfate as clinical evidence does not establish lasting correction of insulin signaling or *beta-cell* function.
    - *Beta cells* store and release insulin.

Vanadyl Sulfate has an insulin-like activity that is capable of imitating insulin actions. Its mechanism of action is through activating the *insulin-receptor* through *tyrosine kinase* and *kinase phosphorylation pathways*. This mechanism is usually accomplished by beta cells; but this is not an established safe treatment for Type 1 Diabetes and *cannot repl*ace insulin.

Proposed benefits (evidence varies)

- Promote proper insulin function
- Promote muscle growth
- Antidiabetic
- Lower blood glucose levels

Possible Side Effects

- Gastrointestinal discomfort
- Gas
- Nausea
- Green tinted tongue
- Diarrhea
- Low energy

Dosages

- No established safe, effective vanadium regimen treats diabetes or builds muscle. High-milligram experimental doses are not routine supplement advice.
- General consumption for the average individual is around 5-20 mcg; this is achieved through normal day to day food consumption

WARNING

Seek medical advice before use. Vanadyl Sulfate may affect people with **Diabetes** as Vanadyl Sulfate can promote and maintain longer insulin levels leading to lower blood sugar levels. There is concern of **kidney damage** resulting from toxicity and high amounts of Vanadyl Sulfate.

Vitamins

Vitamins are organic substances that are needed and vital for growth and proper body function in the human body. Generally, the body doesn't have the capabilities to create enough of them to promote good health and function. As such we receive a good portion of them through diet. Vitamins are either fat-soluble or water-soluble. Fat-soluble means it is absorbed with and stored in fat while water-soluble vitamins dissolve in water and many excess amounts are excreted, although some are stored and high doses can still be harmful. Refer to the end of this section for any applicable warnings.

Vitamin A (Fat-Soluble) **Adult RDA:** **700–900 mcg RAE**

Vitamin A is a group of nutritional compounds needed for proper skin health, immune function, and eye health. R*etinol, retinal* and *retinoic acid are retinoids; beta-carotene is a provitamin A carotenoid that the body can convert to vitamin A. Retinal* is needed by the retina. *Retinal* binds *ops*ins to form light-sensitive pigments: rhodopsin supports low-light vision, while cone pigments support color vision. *Retinoic acid* regulates gene expression involved in cell growth, development and skin health. *Retinol* can be stored as retinyl esters and is a transport form of Vitamin A that can be changed to *retinal* or *retinoic acid* as needed by the body. *Beta-carotene* has been seen as a possible powerful antioxidant as it provides protective effects on the skin against photo-oxidative damage. *Adult RDAs are 700 mcg RAE for women and 900 for men. Excess preformed vitamin A may cause side effects such as peeling skin, diarrhea, vomiting, nausea, vision problems, headaches, and possible yellowing of the skin.*

Vitamin B1 (Water Soluble) **Adult RDA:** **1.1–1.2 mg**

Vitamin B1, also known as Thiamine, plays an important role in energy metabolism. Since the body is not able to synthesize it, it is considered an essential micronutrient. Thiamine and its derivatives play a major role in the breakdown of carbohydrates, *alpha-keto acids (plays a role in the Citric Acid Cycle amongst others)* and amino acids, making use of the carbohydrates to create glucose. *Thiamine diphosphate*, one of its derivatives, also plays a role in the oxidation of glucose for energy. It seems that *Thiamine triphosphate*, a derivative of Thiamine, may have a neurological role as a possible nerve conductor as Thiamine is distributed throughout the nervous system, although currently not enough is known to further state its significance with the nervous system. *Adult RDAs are 1.1 mg for women and 1.2 mg for men. No upper limit is established; this does not prove unlimited doses safe.*

Vitamin B12 (Water Soluble) **Adult RDA:** **2.4 mcg**

Vitamin B12, also known as cobalamin, is an essential vitamin that plays a role in cellular metabolism in practically every cell in the body. It is also required by a few important enzymes; improper function of these enzymes can cause illnesses and diseases. One of the enzymes is *methionine synthase; this* enzyme is involved in DNA synthesis. Another enzyme is *methylmalonyl-CoA mutase; this* enzyme is responsible for energy production/metabolism and can be found in the kidneys, heart, ovaries, brain, muscle, and liver. It can come in a few forms such as *cobalamin, cyanocobalamin, methylcobalamin,* etc. *The adult RDA is 2.4 mcg; pregnancy and lactation needs differ. No upper limit is established because toxicity is low.*

Vitamin B2 (Water Soluble) **Adult RDA:** **1.1–1.3 mg**

Vitamin B2, also known as riboflavin, is an essential vitamin that is required by the body for *cellular respiration*, the process in which your cells convert energy from nutrients into **ATP** and then release the waste molecules. It also plays a part in creating two cofactors necessary for the molecular process. One is *flavin adenine dinucleotide (FAD)*, and the other is *flavin mononucleotide*. Both of these cofactors play a role in *REDOX* functions. *REDOX* is the reduction and oxidation reaction that takes place during the oxidation of molecules like fatty acids or the reduction of oxidative compounds like unstable radicals. They additionally impact lipid metabolism, the breakdown of molecules like fatty acids and cholesterol, once again through *REDOX*. *FAD-dependen*t **glutathione reductase helps maintain reduced glutathione**, a common antioxidant in the body. As glutathione meets an oxidant, glutathione oxidizes to stabilize the oxidant. *This enzyme helps* keep the ratio of oxidized glutathione to glutathione in balance. *Adult RDAs are 1.1 mg for women and 1.3 mg for men. Bright yellow urine is a harmless common effect; no upper limit is established.*

Vitamin B6 (Water Soluble) **Adult RDA:** **1.3–1.7 mg**

Vitamin B6, also known as *pyridoxine,* is a collection of 6 different forms/versions of Vitamin B6. The most commonly known are *pyridoxine, pyridoxal,* and *pyridoxamine.* All of these forms are considered inactive until the body *phosphorylates* them and creates *pyridoxal 5'-phosphate (PLP).* Its major functions come from it acting as an enzyme cofactor. It is essential for amino acid metabolism and the creation of glucose from various sources, carbohydrates and glycogen; the result is extremely beneficial for athletes. Its derivatives also play a role in the creation of neurotransmitters such as *noradrenaline, adrenaline, serotonin, and GABA; this* mechanism can promote and maintain proper emotional health since *PLP* is needed to convert **L-DOPA** into dopamine by the *L-dopa decarboxylase* enzyme. *RDA: 1.3 mg at ages 19–50; older women 1.5 mg, men 1.7 mg. Excess supplements can cause neuropathy. Adult upper limits differ: US 100 mg/day; EFSA 12 mg/day (2023).*

Vitamin C (Water Soluble) **Adult RDA:** **75–90 mg**

The most commonly known vitamin, Vitamin C, can also be called ascorbic acid, and its various derivatives like *l-ascorbic acid,* and *ascorbic acid 2-phosphate* are crucial and needed for healthy function of the human body. It has been seen to be beneficial for skin aesthetics and health as it is involved in **collagen** synthesis. Since *l-ascorbic acid* plays a role as cofactor for *collagen* synthesis, including type I (skin, bone, tendon), type II (cartilage) and *type III (soft tissues),* it's essential in the maintenance and development of connective tissues like cartilage and tendons. It can also aid with oxidative damage by sequestering free radicals that can damage cells in the body. Additionally, ascorbic acid plays a role as a cofactor for the enzyme *dopamine-β-hydroxylase* that converts *dopamine* into *noradrenaline.* It is also able to turn cholesterol into bile acids by modulating *7α-hydroxylation* which limits the amount of cholesterol that can be broken down; proper function can thereby reduce cholesterol levels. Vitamin C is not capable of preventing a cold; however, it may shorten and lessen the symptoms as it can enhance *T-cell proliferation (white cell multiplication)* in the presence of an infection. *Adult RDA: women 75 mg, men 90 mg; smokers need 35 mg more. Excess may cause diarrhea, cramps and nausea.*

Vitamin D (Fat-Soluble) **Adult RDA:** **600–800 IU**

Vitamin D is a group of "broken" steroids as one of its carbon chains is broken. It is considered an essential vitamin and can be synthesized in the body through cholesterol and sun exposure with enough UV on the skin; it can also be found in fish and eggs. There are a total of 5 Vitamin D forms, but only Vitamin D2 *(ergocalciferol)* and Vitamin D3 *(cholecalciferol)* are used in the body, while both can raise blood levels; D3 often raises and maintains them more effectively. D3 occurs in oily fish and fortified foods; dairy content depends on **fortificatio**n. D2 occurs in UV-exposed fungi. Adult RDA: 600 IU (15 mcg) through 70, then 800 IU (20 mcg). Vitamin D plays a role in the modulation of cellular *proliferation (multiplication)* or *apoptosis (programmed cell death),* promoting insulin production in *beta cells (cells that store and release insulin).* It is also able to promote protein synthesis through the *Vitamin D Receptor* in the nucleus of the cell. Vitamin

D3 can also positively modulate the enzyme *aromatase (produces estrogen)* by either increasing or decreasing the activity, dependent on the tissues/cells. Vitamin D has also been seen to protect neuronal death from *excitotoxicity (nerve cells killed from overstimulation)* by positively modulating a large set of calcium channels by down-regulating channels that are sensitive to *Calcium Ions,* like the *L-type calcium channel* which has a long-lasting effect after activation. It is also able to reduce *atherogenesis (adhesion of white blood cells to the surface of blood cells).* These cell-level mechanisms do not establish cardiovascular protection from supplements; foam cells arise from lipid-loaded macrophages in vessel walls. Vitamin D supports calcium and phosphate availability for bone mineralization; deficiency impairs bone health. *Higher than recommended doses may promote side effects such as metallic taste, headache, dry mouth, nausea, vomiting, fatigue, and weakness.*

Vitamin E (Fat-Soluble) **Adult RDA:** **15 mg**

Vitamin E is a group of 8 compounds, and although the full mechanism of Vitamin E is not fully understood, it is believed that it does aid antioxidant processes in the body. It is important to note that like many fat-soluble vitamins, this one more so, it is not recommended to take in high doses as it may cause damage to the body. A balanced diet is generally enough to receive the daily recommended amount; supplementation is usually not necessary, and the adult RDA is 15 mg of alpha-tocopherol. High-dose supplements can increase bleeding and other harms; 400 IU is not a universal fatal threshold. Vitamin E is divided into two categories: one is tocotrienols, the other *tocopherol, alpha tocopherol* being the most active. Vitamin E participates in an antioxidant network with Vitamin C and other nutrients; it does not require a combined supplement to work. The variations of *tocopherols* have different antioxidant mechanisms; alpha *tocopherol* stops the oxidation and creation of new radicals like *reactive oxygen* species, while *tocopherol gamma* neutralizes current radicals. Since Vitamin E is a lipid-soluble nutrient, it is able to embed itself in the cell's membrane; this provides maximum first-line protection against oxidation of the cell's membrane and lipid oxidation. This can provide benefit body-wide as it is able to promote cell repair which could enhance muscle growth or just general well-being by prolonging cell life. To naturally include more Vitamin E in your diet, almonds and avocados would be a good choice as they are high in Vitamin E. *Higher than recommended doses may promote side effects such as headache, nausea, vomiting, diarrhea, fatigue, bruising and bleeding.*

Vitamin K (Fat-Soluble) **Adult AI:** **90–120 mcg**

Vitamin K is generally referred to as a group of various Vitamin K compounds, mainly Vitamin K1 *(phytomenadione/phytonadione),* Vitamin K2 *(menaquinone),* broken down further into MK-4 *(menaquinone-4)* and MK-7 *(menaquinone-7).* Vitamin K is required by the body for one's health and complete synthesis of some proteins involved in blood coagulation. It's important to note that Vitamin K3 is a synthetic version of Vitamin K and has been seen to be toxic. Also, it's important to note the difference between the element K, which is potassium, and Vitamin K. They are not the same thing and are not interchangeable; Potassium is a mineral while Vitamin K is not. We generally consume Vitamin K1 from a diet rich in green leafy vegetables and Vitamin K2 from fermented foods, cheeses, eggs and butter. *Gamma-carboxyglutamate* is created during the Vitamin K cycle, and this is where some of its main effects come from. Like glutamate, *gamma-carboxyglutamate* can attach itself to calcium which can lead to de-calcification of soft tissues like the heart and arteries by acting on *Matrix Gla Protein; this* protein is involved in de-calcifying soft tissues. Fortunately, it seems to have the opposite effect on bone promoting mineral accumulation in bone. It also seems to positively influence *mitochondrial* **ATP** production which could promote more energy in the body. *Adult adequate intakes are 90 mcg for women and 120 mcg for men. Supplement trials have not established arterial decalcification or an energy boost; warfarin requires consistent intake.*

VITAMIN WARNINGS

Seek medical advice before use. There is concern that…

Vitamin A

Vitamin A may affect those with **Type V hyperlipoproteinemia** as they can increase the chance of Vitamin A poisoning as the increase of lipids in the blood can increase Vitamin A absorption. Vitamin A can also make **liver disease** worse as Vitamin A is stored in the liver.

Vitamin B1

There currently is no known adverse reaction with Vitamin B1 and other diseases or illnesses.

Vitamin B12

Vitamin B12 can cause allergic reactions in people sensitive to cobalt. Cyanocobalamin requires specialist advice in **Leber hered**itary optic neuropathy; do not generalize this concern to all forms and all eye disease.

Vitamin B2

There currently is no known adverse reaction with Vitamin B2 and other diseases or illnesses. Some medications may increase or decrease absorption, but currently the fluctuation amounts are not clear if this is a concern.

Vitamin B6

Prolonged supplemental Vitamin B6 can cause sensory nerve damage. Risk depends on dose and duration; use the applicable upper limit rather than assuming water-soluble vitamins are harmless.

Vitamin C

High-dose Vitamin C can cause gastrointestinal symptoms and may increase urinary oxalate; seek advice with kidney disease, stone history or **iron overload.** It can interfere with some glucose *tests*. Routine dietary intake is not established to worsen diabetes or schizophrenia; treatment combinations *need individual revi*ew.

Vitamin D

Vitamin D can increase the risk of kidney stones or higher than normal levels of calcium in people with **Histoplasmosis**, **sarcoidosis, tuberculosis**, and **lymphoma** as Vitamin D can increase calcium levels.

Vitamin E

High-dose Vitamin E supplements can increase bleeding risk, including with antic**oagulants, and** have not been shown to prevent **cardiovascular disease**. A large trial found increased prostate-cancer risk with 400 IU/day synthetic Vitamin E in men. Disease-specific **hig**h-dose t**reatment requires** medical advi**ce**.

Vitamin K

Vitamin K can counteract warfarin. Maintain a consistent intake and arrange clinician-guided monitoring when changing diet or supplements. Avoiding all Vitamin K is unnecessary. Vitamin K supplements are not established treatment for arterial calcification, diabetes or heart disease.

Waxy Maize Starch

Waxy maize, in regard to the name, is technically waxy corn. The actual supplement, on the other hand, is a reference towards fast-digesting carbs derived from various starchy foods like rice, corn, barley, etc. More so gearing towards the water-soluble *polysaccharide (complex carbohydrate)* glucose molecule known as *amylopectin.* It has generally been seen in some, and at times, marketed separately as a post-workout supplement with the attempt to restore glycogen levels and currently has been looked upon as a potential aid for those who have *glycogen storage diseases*.

The general belief and marketing schemes have been that *amylopectin* is a superior fast-acting carb with claims that state it is able to be absorbed and replenish glycogen levels 70-80% faster than other fast carbs like *maltodextrin* and *dextrose*. However, it seems that unrefined raw Waxy Maize has such slow digestibility that it can be considered a low glycemic food. Heat-treated or cooked Waxy Maize has near equivalent properties to that of *dextrose* and *maltodextrin* although not 100% equivalent. However, it is important to note that, although it may be inferior to other fast carbs, the intake of Waxy Maize would still replenish glycogen levels compared to no intake of any carbs after exercise.

In regards to *glycogen storage diseases, it* seems <u>the main reason that it could be beneficial for this disease is the exact opposite of how it's marketed, and it's purported sold use, as a slower digesting longer-lasting carbohydrate</u> that can promote *normoglycemia; this* is recognized as longer periods of consistent, normal, blood glucose levels, over a longer period of time. Heat-treated Waxy Maize starch has shown benefit in a favorable glucose profile in those suffering from *glycogen storage disease.*

- Heat modified/treated Waxy Maize has been looked upon as it is easier to digest and not as irritable to the intestines as uncooked raw Waxy Maize or other starches. Since raw starches tend to cause bloating, gastrointestinal discomfort, excessive gas, and diarrhea. Heat-treated/modified Waxy Maize can drastically reduce these symptoms.

<u>Proposed benefits (evidence varies)</u>

- Low Glycemic carb
- Slow Digesting carb
- May promote longer glucose levels
- May aid in Glycogen Storage Disease

<u>Possible Side Effects</u>

- Gastrointestinal discomfort
- Gas
- Diarrhea
- Acne
- Bloating

<u>Dosage</u>

- *Glycogen-storage-disease starch regimens require individualized medical nutrition therapy; a fixed 50 g dose is not suitable for all patients or products.*

WARNING

Seek medical advice before use. Waxy Maize may affect people with **diabetes** as Waxy Maize can affect insulin and blood glucose levels. Waxy Maize has been known to reduce insulin and early insulin response and as a result, may increase blood sugar levels for a longer period of time.

Wheat Germ

First things first, on the off chance that this mistake will be made, it is important to note that Wheat Germ and Wheat Germ Oil are two different types of products. The oil being more of a supplement while Wheat Germ is, in a sense, a grain or food. In wheat germ oil there is the active compound *octacosanol*, a fatty acid alcohol that is claimed to increase muscle and lower body fat. On the other hand, Wheat Germ as a food product is claimed to be a complete grain with a high nutrient profile when compared to all other grains, and like other grains it has been seen to help reduce cholesterol.

Nutrition Content	**1 cup of Wheat Germ**	**Nutrition Content**	**1 cup of Wheat Germ**
Protein	- 33 g	Phosphorus	- 1295 mg
Carbohydrates	- 5 g	Magnesium	- 363 mg
Monounsaturated Fat	- 10 g	Calcium	- 7.2 mg
Saturated Fat	- 2 g	Iron	- 10.5 mg
Fiber	- 20 g	Potassium	- 1070 mg
Vitamin B1	- 1.1 mg	Manganese	- 22.6 mg
Vitamin C	- 6.8 mg	Sodium	- 4.5 mg
Vitamin E	- 18.1 mg	Copper	- .7 mg
Riboflavin	- .9 mg	Selenium	- 73.5 mg
Niacin	- 6.3 mg	Folate	- 400 mcg
Thiamin	- 1.9 mg	Pantothenic Acid	- 1.6 mg

Wheat Germ oil, on the other hand, has a higher *octacosanol* content than wheat germ alone, (which I would like to add can be found in larger concentrations in beeswax); this may potentially enhance athletic abilities; it is believed to accomplish this by enhancing the use of glycogen in muscle tissue while promoting higher levels of the antioxidant **glutathione** in the body.

- Glutathione is produced in the body; it's non-enzymatic and regulates oxidative damage in the body. It is able to stabilize free radicals by donating an electron to oxidants.

Benefits

- Good source of nutrients
- May promote lean muscle
- Promote antioxidant activity
- May promote fat loss
- Promote feelings of fullness
- Improve endurance

Possible Side Effects

- Diarrhea
- Soft stools
- Loss of appetite
- Constipation
- Nausea
- Gas

Dosages

- 1 cup of wheat germ (contains 430 calories)

WARNING

Please consult with a medical professional prior to administering Wheat Germ. There is concern that Wheat Germ may affect those diagnosed with **immune disorders** as Wheat Germ may promote immune system function and worsen symptoms.

Wheat Grass

Wheat grass is the first leaves that have sprouted from the common wheat plant, *T. aestivum*. It has been grown and sold for years as a health and dietary supplement with various health claims, many of which have no merit. However, that is not to say Wheat Grass does not have any benefit; Wheat Grass does contain *chlorophyll*, various amino acids, vitamins and minerals and although it is a different part of the wheat plant from **Wheat Germ**, it does contain *octacosanol*, all of which contribute to its health properties. Wheat Grass has a light sweetened-like taste and can be consumed as juice, tablets or as powder mixes; it can also be seen as a topical applicant in the form of lotions and gels. (If wheat grass juice has a bitter taste it's more than likely spoiled, DO NOT CONSUME!)

Nutrition Content	1 oz Wheat Grass Juice	Nutrition Content	1 oz Wheat Grass Juice
Protein	- 860 mg	Phosphorus	- 21 mg
Beta-carotene	- 120 IU	Magnesium	- 8 mg
Vitamin B12	- .30 mcg	Calcium	- 7.2 mg
Vitamin C	- 1 mg	Iron	- .66 mg
Vitamin E	- 880 mcg	Potassium	- 42 mg

Chlorophyll is the pigment plants use to capture light energy. It shares a ring framework with heme, but contains magnesium rather than heme's iron and is not an oxygen-carrying protein. Laboratory studies of chlorophyll derivatives explore carcinogen binding, oxidative signaling and cancer-cell growth. These findings do not mean that consumed chlorophyll replaces *hemoglobin*, oxygenates tumors or treats cancer. Cancer cells can also use oxygen; their altered metabolism is more complex than an exclusively anaerobic state. Clinical anticancer benefit from th*is supplement is unestablished*. Wheat Grass also contains *Superoxide Dismutase (SOD); oral enzyme content does not establish systemic anticancer benefit.*

- *SOD* is an antioxidant that is able to convert the *Superoxide Radical (O_2^-)* into oxygen and hydrogen peroxide; Superoxide *Radicals* are toxic and can cause oxidative damage in the body.

Proposed benefits (evidence varies)

- Good source of nutrient
- Anti-cancerous
- Antioxidant

Possible Side Effects

- Nausea
- Vomiting
- Diarrhea
- Loss of appetite
- Constipation

Dosage

- Juice serving sizes vary; 1 fluid ounce is about 30 mL, not one cup. No established disease-treatment dose exists.
- For pill or powder form adhere to label instructions

WARNING

Seek medical advice before use. Wheat Grass may affect people with **diabetes** as Wheatgrass can lower blood sugar levels; medications may need to be modified.

White Kidney Bean *(Phaseolus vulgaris)*

White Kidney beans are just regular kidney beans but with a white appearance. They are Native to parts of South America and Europe. The most researched and marketed benefit of White Kidney Bean extract is for its ability to effectively block carbohydrate absorption in the body through the use of *glycoproteins (proteins with carbohydrate chains attached to amino acid chains).*

White Kidney Beans have been seen to reduce carbohydrate absorption by inhibiting enzymes that absorb and break down carbohydrates into glucose. Its function is the inhibition of the *alpha-amylase enzyme. The alpha-amylase enzyme* is responsible for starch/carbohydrate digestion. Inhibition of the *alpha-amylase enzyme* is accomplished through the main bioactive component in White Kidney Bean extract, *alpha-AI1 isoform. Alpha-AI1 isoform* has been seen to have anti-amylase activity as it is able to completely block access to the active site of the *alpha-amylase enzyme.*

- The result of this mechanism is the passing of undigested carbohydrates through the colon, which can either be excreted or fermented by bacteria. If fermented by bacteria, it can result in some of the possible side effects listed below, more so gas than any other side effect.
    - When some of these carbohydrates become fermented by bacteria it is believed that the end result is short-chain fatty acids, particularly *butyric acid, acetic acid,* and *propionic acid.* It is believed these fatty acids are involved in the reduction of colon cancer risk although current data to support this is preliminary.

IMPORTANT KEYNOTE: Acquire extract through a reputable company as raw kidney beans contain a toxic compound known as *phytohaemagglutinin (PHA); this* chemical has been linked with severe gastrointestinal effects in humans. Generally, this chemical can be removed through cooking and boiling; additionally, poor filtration or extract processes in subpar products can leave copious amounts of *PHA* while having negligible amounts of *anti-amylase* activity. The product known as Phase 2 has the highest purity and filtered white kidney extract with a filtration process that is able to deactivate PHA, although there are other high-quality products; Phase 2 has gone through extensive testing.

Benefits

- Aid fat loss
- Promote weight loss
- Carb blocker
- Possible antidiabetic

Possible Side Effects

- Gastrointestinal discomfort
- Upset stomach
- Vomiting
- Nausea
- Diarrhea
- Gas

Dosage

- 3 g of White Kidney Bean extract broken up amongst every carbohydrate-rich meal throughout the day.

WARNING

Please consult with a medical professional prior to administering White Kidney Bean extract. There is concern that White Kidney Bean extract may affect those who have been diagnosed with **Diabetes** as White Kidney Bean extract has been known to lower blood sugar, as it can reduce carbohydrate absorption in the body, leading to lower blood sugar levels; medications may need to be modified.

White Willow *(Salix alba)*

The White Willow is from Willow family *Salicaceae; it* is a medium to large-sized tree; its foliage is long green leaves with a white-ish underside giving its name of white willow. It has a history of use in traditional medicine in China, Europe, Ancient Greece; it even has had documentation of use going as far back to the Sumerian and Assyrian civilization. It was believed that it could reduce pains, aches, and fevers. The White Willow tree is native to Europe and parts of Western Asia. White Willow can still be found in some products and it is a natural alternative to synthetic versions of the product such as Aspirin.

The bioactive compound and extract of the White Willow tree is salicin; if it sounds familiar it's because you have heard of its chemical derivatives like Aspirin *(Acetylsalicylic Acid)*, as mentioned before, and the commonly used acne medication *Salicylic Acid*. *Salicin* has been seen as a potent anti-inflammatory and as an antioxidant but it can still cause gastrointestinal irritation, allergy *and salicylate-related bleeding; it is not risk-fr*ee aspirin.

- *Salicin* has been seen to promote antioxidant activity by enhanced *REDOX homeostasis (maintained stability).*
    - Stability is especially crucial for REDOX; REDOX stands for *Reduction* and *Oxidation*. *Oxidation* is where an atom loses an electron and oxidizes, while *Reduction* is where an atom gains an electron and becomes less oxidative. This is the basis for either oxidant and oxidative damage or antioxidant activity, i.e. the reduction/stabilizing of oxidative molecules in the body. As you can see *REDOX* stability is crucial for antioxidant and metabolic effects in the body.
- *Salicin* has been seen to provide an anti-inflammatory effect through similar fashion as aspirin. *Salicin* converts into *Salicylic Acid* in the liver. Like aspirin, *Salicylic Acid* is then able to inhibit *Cyclooxygenase (COX) 1* & *2*.
    - *COX 1* & *2* are enzymes that are involved in the formation of prostanoids; these are fatty acid/lipid inflammatory mediating molecules like Arachidonic Acid.

Proposed benefits (evidence varies)

- Anti-inflammatory
- Antioxidant
- Fever Reducer
- Analgesic (Pain reliever)

Possible Side Effects

- Itching
- Upset stomach
- Rash

Dosage

- Some standardized extracts supply 120–240 mg salicin/day; this is not the same as 240 mg of whole bark. Follow the product and medical advice.

WARNING

Seek medical advice before use. White Willow may affect people with **Peptic Ulcer Disease**, **Renal Disorders,** and **Liver Disease** as it is metabolized in the liver and has some activity in the kidneys much like aspirin. **Children** should not be given White Willow or its extract as there is a risk of developing Reye's **Syndrome**. Willow Bark can also increase the risk of bleeding in those with **Bleeding Disorders.**

Wild Oat (*Avena sativa/Neuravena ®*)

Avena sativa is part of the *Poaceae (grass)* family and can be seen referred to as oats or wild oats. It is native to Europe but has been domesticated in the Middle East and has been generally looked upon as a weed. The parts used are the stems and seeds that can be rolled into oats as seen in oatmeal; consumption does not provide the same health benefits as the extract. Recently an extract made from wild green oats under the name Neuravena® has come out with some health benefits gearing towards cognitive health. Each, the oats and extract, have separate and shared benefits. Small studies concern specific green-oat extracts; ordinary oats and extracts are not interchangeable or established treatments for addiction, depression or low testosterone.

A tincture of *Avena sativa* fresh leaves has been used in India to help addictions and may help counteract nicotine withdrawals. As it turns out there may be truth to this as the tincture may have the bioactive compounds *avenanthramides (this is specific to oats)* which have been seen to interact with *dopamine* responses. Neuravena® is an extract based on the isolated bioactive compounds. Although the full mechanism and function are not fully known, it is believed that its main mechanism is through the inhibition of two enzymes, *monoamine oxidase-B (MAO-B)* and *phosphodiesterase 4 (PDE4).*

- *MAO-B* is responsible for breaking down various amines, in interest *phenylethylamine (involved in mood, attention, and energy)* and *dopamine (involved in reward-motivation behavior/neurotransmission).* Higher levels of these amines mean higher levels of cognitive function and fewer depressive episodes.
- *PDE4, in general,* is involved in signal regulation as it breaks down *cyclic nucleotides; these* are molecules that are *intracellular (inside the cell)* messengers that respond to *extracellular (outside the cell)* signals. Inhibition of the variations of *PDE4 (A, B, C,* and *D)* has been seen to have antipsychotic and antidepressant effects as well as a reduction of various other dysfunctions in the body.

As food, oats have been shown to have a great number of nutrients. As an additive to products, it has been shown to aid with things like skin hydration as the oats' straws can be used in bath products to improve skin health and may help individuals that suffer from eczema. It has also been added to proteins for use as a mass gainer; the calories and nutrients can support weight gain, but a reliable pituitary-mediated testosterone increase is not established.

Proposed benefits (evidence varies)

- Promote feelings of well-being
- Mass gainer
- Improve skin quality
- Possible cognitive aid
- Aid with withdrawals

Side Effects

- Increased heart rate (Extract side effects)
- Sleeplessness (Extract side effects)
- Restlessness (Extract side effects)
- Bloating (When eaten)
- Gas (When eaten)
- Upset Stomach (For both)

Dosage

- Recommended 200 g of oats a day, for the extract adhere to label instructions

WARNING

Seek medical advice before use. There currently is a lack of data on the extract; however, if you're **allergic to oats, Avena** sativa may cause an allergic reaction.

Yerba Mate *(Ilex paraguariensis)*

Yerba mate is a tree that is indigenous to South America; Yerba matte starts out as a shrub that eventually becomes a tree; the tree can grow upwards to 15 meters with evergreen leaves. The leaves are where most of the bioactive compounds are at, and can differ in potency and levels dependent on whether the tree is a female or male. It has been advertised and marketed as an appetite suppressant, fat burner, an antioxidant and anti-inflammatory; this can provide benefit to the cardiovascular system.

Yerba Mate benefits mainly come from various compounds found in the plant, mainly *polyphenols* and *xanthines*. *Polyphenols* can be seen as where antioxidant activity derives from in plants (Not all *phenols* are good; some can be toxic). You probably interact with *xanthines every day: the caffeine* found in your coffee or the theobromine in chocolate; these are *xanthines.* Yerba Mate has the *xanthines* caffeine, theobromine and theophylline and the *polyphenol* **quercetin**. It is recommended to acquire a crude or raw form as processing can destroy these components.

- Its appetite and fat burning capabilities come from its caffeine content as caffeine has been seen to increase (GLP-1) activity; GLP-1 directly inhibits gastric emptying and stomach acid expression thereby decreasing appetite, reducing calorie intake and promoting fat loss.
- Theobromine can inhibit *phosphodiesterase* enzymes; *phosphodiesterase* enzymes can convert *cAMP* into an inactive form. *cAMP* has a role as a messenger of hormone and neurotransmitter-controlled systems, involving mood regulation all the way to inflammation and vasodilation. This general mechanism can promote vasodilation and feelings of well-being in the body.
- Theophylline has been used in therapy for respiratory diseases like asthma as it's a bronchodilator through inhibition of *phosphodiesterase 3 & 4* and its anti-inflammatory effect seems to come from this *histone deacetylase activation; this* turns off inflammatory genes. Theophylline blocks *Adenosine receptors; both* of the mechanisms are derived from there.
- Quercetin can enhance anti-inflammatory effects through *heme-oxygenase-1 production; this* can directly inhibit nitric oxide and inflammatory markers released in response to inflammation.

Proposed benefits (evidence varies)

- Bronchodilator
- Anti-inflammatory
- Antioxidant

Possible Side Effects

- Gastrointestinal discomfort
- Anxiety
- Restlessness
- Nausea
- Nervousness
- Vomiting
- High Blood Pressure
- Headache
- Arrhythmia

Dosage

- 1-1.5 g of Yerba Mate in powdered form or tea

WARNING

Seek medical advice before use. Yerba Matte may affect those who have caffeine sensitive conditions such as **High Blood Pressure**, **Glaucoma**, **Anxiety,** and **Heart Problems**. IARC classifies drinking very hot beverages (above 65°C) as probably carcinogenic. Mate that is not very hot was not **classifiable; smok**ing, **alcohol** and preparation can affect risk. Yerba Mate can increase low blood sugar symptoms in **Diabetics**.

Zinc

Zinc is a post-transitional metal that can be found on the periodic table under the atomic number 30 and is one of the most abundant elements in the earth's crust. It is also an essential element that's involved in various enzymes within the body and has been seen to be an immune booster and antioxidant; additionally, it has been used in various supplements to boost testosterone levels. It can be found in eggs, meat, some seafood products like shellfish and legumes; still, the relative problem with the majority of people is poor bioavailability of Zinc since it has poor absorption.

In Zinc deficiencies, *Interleukin 1β (IL-1β)*, an inflammatory and fever response protein, has higher production than those with normal Zinc levels. Zinc supplementation is able to reduce anti-inflammatory signals by up-regulating *A20* mRNA which inhibits cell-specific *NF-kB*, since *IL-1β* is an activator of NF-kB (NF-kB activates genes involved in inflammation); this halts the signaling chain, reducing the abnormal expression of inflammatory *cytokines (cell messengers)*. Additionally, zinc is a co-factor of the enzyme *Super Oxide Dismutase (SOD); NADPH oxidase* contributes a great deal of free radicals in the body; zinc is able to inhibit this enzyme and aid in oxidant reduction.

- *SOD* is able to convert superoxide into oxygen and hydrogen peroxide.

Zinc is also able to enhance immune functions in the same manner as mentioned before since it is able to up-regulate *A20 mRNA (a direct cell-specific inhibitor of NF-kB)*. Like mentioned before, this mechanism can decrease inflammatory gene expression in *IL-1β* as well as *IL-8 (directs immune cells target)* and *Tumor Necrosis Factor-α (TNF-α)*.

- *TNF-α* is a *cytokine that* regulates immune cells; dysregulated *TNF-α* has been linked to various disease states, such as cancer, arthritis, Alzheimer's and psoriasis.

Zinc deficiency can contribute to low testosterone, but it is not established as the cause in most men, as zinc plays a role in testosterone synthesis. Androgen receptors have a Zinc binding site, which may be the reason as to why testosterone levels fluctuate according to Zinc levels; proper Zinc levels can maintain normal resting Testosterone levels; however, it will not increase it above baseline; excess Zinc Supplementation can cause oxidative damage, although it can temporarily increase testosterone levels after exercise.

Benefits

- Testosterone balance
- Anti-inflammatory
- Antioxidant
- Rheumatoid Arthritis aid
- Immune enhancer

Possible Side Effects

- Metallic taste
- Kidney damage
- Stomach damage
- Diarrhea
- Nausea
- Vomiting
- Fever
- Upset stomach
- Fatigue

Dosage (Dosages Vary dependent on variation, Zinc Citrate vs. Zinc Sulfate; adhere to label directions)

- Adult RDA: women 8 mg/day, men 11 mg/day. Adult upper limit: 40 mg/day total. Deficiency treatment needs individual dosing; prolonged excess can cause copper deficiency.

WARNING

Seek medical advice before use. Excess Zinc can impair immunity and cause copper deficiency. With **HIV/AIDS,** assess deficiency and medicines individually; routine nutritional intake is not established to accelerate disease. Monitor glucose if using diabetes medicines.

<u>ZMA *(Zinc Monomethionine Aspartate)*</u>

ZMA contains ***Z****inc monomethionine, Magnesium aspartate* and Vitamin B6. ZMA has been marketed like an anabolic mineral supplement, more so for support of muscle recovery than a steroid-like product. Correcting deficiency can help normal function; trials do not reliably increase testosterone or training adaptations in adequately nourished athletes. Sweat losses can contribute to mineral needs, but deficiency should not be assumed in most athletes. ZMA supplementation can improve deficient levels, indirectly improving other areas. Vitamin B6 has independent coenzyme roles; added B6 does not establish improved magnesium absorption or ZMA performance.

The enzyme *NADPH oxidase* contributes a great number of free radicals in the body; zinc is able to inhibit this enzyme. With the addition that Zinc is a co-factor for the enzyme *Super Oxide Dismutase (SOD), SOD* is able to convert superoxide into oxygen and hydrogen peroxide. Both of these mechanisms can greatly either reduce the amount of radicals created, or stabilize radicals in the body thereby reducing oxidative damage that comes as a by-product of energy created during strenuous exercise improving recovery time. Zinc has been linked to maintaining healthy baseline testosterone in deficient individuals.

Magnesium is found in the body as magnesium ions; 25% is found in muscle tissue, and 60% is found in our bones. Magnesium plays an important role in **ATP *(Adenosine Triphosphate)*** metabolism. The *ATPases* are enzymes that are dependent on calcium and magnesium ions. *ATPase* breaks down ATP into *ADP (Adenosine diphosphate).* This *dephosphorylation (removal of a phosphate group)* is used to create energy. Proper magnesium levels become extremely vital for any athletic event or sport; during strenuous activity the body can break down and reform 529 oz of ATP, even though the body can only store 3 oz at one time.

- Increased magnesium intake, <u>*if one is lacking*</u>, can lead to increased metabolic efficiency, optimal heart rate, and oxygen consumption. The result is an increased aerobic capability.

Aspartate has been seen to act as a neurotransmitter and stimulate neurotransmitters which could enhance cognitive function. Additionally, it can influence the release of *luteinizing hormone* which directly induces testosterone synthesis and may increase overall testosterone in the body.

<u>**Benefits**</u>

- Rheumatoid Arthritis aid
- Testosterone balance
- Anti-inflammatory
- Immune enhancer
- Antioxidant

<u>**Possible Side Effects**</u>

- Diarrhea
- Nausea
- Vomiting
- Fever
- Upset stomach
- Fatigue

<u>**Dosage**</u> (Dosages Vary dependent on variation, Zinc Citrate vs. Zinc Sulfate; adhere to label directions)

- 450 mg supplemental Magnesium exceeds the US adult 350 mg/day limit. Count all sources of Zinc and B6.

<u>***WARNING***</u>

Seek medical advice before use. Review the Zinc, Magnesium, Aspartate and Vitamin B6 entries for additional risks. Magnesium aspartate is not D-Aspartic Acid.

Alternative Names Index

1,2,3-propanetriol 89
1,3,7-trimethylxanthine 34
3,3′-dihydroxy-β,
β′-carotene-4,4′-dione 17
5,7- dihydroxyflavone 44
3, 3’diindolylmethane 62
3,5,4’- trihydroxystilbene 170
3β-Hydroxy-5-Androstene-17-one 59
3,7-dimethylxanthine 186
4-hydroxyphenylalanine 125
5-methyl-7-methoxyisoflavone 138
7-isopropoxyisoflavone 107
Actaea racemosa 27
Allium sativum 81
Alpha-hydroxy-isocaproic Acid 100
Ananas comosus 31
Angelica gigas 57
Angelica sinensis 64
Arctium lappa 32
Arthrospira platensis 180
Ascophyllum nodosum 110
Astragalus membranaceus 18
Astragalus propinquus 18
Avena sativa 204
Beta-alanyl-L-histidine 35
Beta-Hydroxy beta-methylbutyrate 101
Bitter Melon 141
Bitter Orange 185
Borago officinalis 28
Capsaicin 37
Camellia sinensis 95
Centella asiatica 92
Cimicifuga racemosa 27
cis-5,8,11,14-eicosatetraenoic acid 14
Citrus aurantium 185
Cola nitida 111
Coleus Forskohlii 74
Cocoa 186
Crataegus pinnatifida 99
Cynara scolymus 15
Diferuloylmethane 55
Devil’s Weed 187
Dog Rose 173
Epimedium 104
Eurycoma longifolia jack 128
Euterpe oleracea 6
Euphausia superba 112
Evodia rutaecarpa 69
Forskolin 74
Garcinia mangostana 136
Glycine propionyl-L-carnitine 78
Glycyrrhiza glabra 127
Hordeum vulgare 21
Hydrastis canadensis 91
Hypericum perforatum 181
Ilex paraguariensis 205
Irvingia gabonensis 9
Indian saffron 189
L-Alanyl-L-Glutamine 184
Lactobacillus Acidophilus 8
Lagerstroemia speciosa 20
Lepidium meyenii 133
Linseed Oil 72
Linum usitatissimum 72
Lycium barbarum 90
Medicago sativa 11
Methylsulfonylmethane 132
Morinda citrifolia 145
Monacolin K 169
myo-Inositol 106
N-dimethyltyramine 103
Neuravena 204
Oenothera biennis 68
Paullinia cupana 96
Phaseolus vulgaris 202
Phosphatidylcholine 126
Pinebark extract 164
Plectranthus barbatus 74
Prunus africana 165
Punica granatum 156
Rosa canina 173
Salai guggul 30
Salix alba 203
Salvia hispanica 38
Sambucus nigra 67
Serenoa repens 176
Simmondsia chinensis 109
Silybum marianum 140
Taraxacum officinale 56
Tetradium ruticarpum 69
Trigonella foenum-graecum 70
Trimethylglycine 23
Uncaria tomentosa 36
Undaria pinnatifida 75

Alternative Names Index

Vaccinium myrtillus 24
Valeriana officinalis 193
Vitis vinifera 93
Wakame 75
Withania somnifera 16
Zingiber officinale 82

Made in United States
Orlando, FL
04 November 2026

90204179R00127